W9-ALW-113

With Respect for Nature

SUNY series in Environmental Philosophy and Ethics

J. Baird Callicott and John van Buren, editors

With Respect for Nature

Living as Part of the Natural World

J. Claude Evans

State University of New York Press

Published by
State University of New York Press, Albany

Printed in the United States of America

For information, address State University of New York Press,
194 Washington Avenue, Suite 305, Albany, NY 12210-2365

Production by Kelli Williams
Marketing by Susan M. Petrie

Library of Congress Cataloging-in-Publication Data

Evans, Joseph Claude.
With respect for nature : living as part of the natural world / J. Claude Evans.
p. cm. — (SUNY series in environmental philosophy and ethics)
Includes bibliographical references and index.
IBSN 0-7914-6443-1 (hardcover : alk. paper) — ISBN 0-7914-6444-X (pbk. : alk. paper)
1. Animal rights. 2. Animal welfare—Moral and ethical aspects. 3. Hunting.
4. Respect. I. Title. II. Series.
HV4708.E93 2005
179'.3—dc22

2004017709

10 9 8 7 6 5 4 3 2 1

Contents

Preface

Many environmentalists say that unless we change the way we live, we will destroy both the world we live in and ultimately ourselves. Fundamental changes are required in our understanding both of ourselves and of the natural world. Environmental thinkers have also argued that the sources of environmental crisis lie deep in the origins of Western civilization, in our most fundamental understanding of what it is to be human and to live an appropriately human life, and in our conception of the natural world. If this is true, then in addition to deciding how to regulate our industry—controlling pollution, economizing on resource use, developing sustainable agriculture, etc.—we face problems that call for a fundamental moral reorientation.

In all religious and philosophical traditions, one's sense of moral obligation—be it to other human beings, to animals, or to nature—always involves a specific understanding of both the moral self and the other to whom the self relates in moral action. All the major ethical theories in the Western philosophical tradition—from Socrates and Plato, through Aristotelian virtue ethics and Kantian deontology, to utilitarianism—have disagreed about two related ideas: what it means to be human and what it means to relate morally to an other. However, all of these traditions have shared the basic assumption that the morally significant other is human: family, friends, fellow citizens, or fellow human beings. This is true even when these moral relationships are established or supported by God. Given this tradition, any deep shift in our relationship to nature will require a new understanding of what it means for human beings to relate to the natural world and to live a moral life as part of that world.

TRADITIONAL ANTHROPOCENTRISM

In most of the Western religious and philosophical tradition, the nonhuman world is thought to exist for the sake of human beings. This metaphysical and ethical position has come to be known as *anthropocentrism.*[1] It is based on religious doctrine, on philosophical argument, and on scientific theory.

The assumed superiority of human beings over the rest of the world has religious expression in the first account of creation in *Genesis*. God commands human beings, who have the unique status of being created in the image of God, to "be masters of . . . all the wild beasts" (*Genesis* 1:26) and to "fill the earth and conquer it" (*Genesis* 1:28).[2] Independently of the Judeo-Christian religious tradition, there are statements of anthropocentrism in the Western philosophical tradition beginning with the ancient Greeks. Xenophon formulated the classic position in his dialogue, *Memorabilia,* in which his "Socrates" says,

> *Tell me, Euthydemus, has it ever occurred to you to reflect on the care the gods have taken to furnish man with what he needs? . . . Now, seeing that we need food, think how they make the earth to yield it, and provide to that end appropriate seasons which furnish in abundance the diverse things that minister not only to our wants but to our enjoyment . . . and is it not evident that they [the lower animals] too receive life and food for the sake of man?* (Xenophon, Book IV, Chapter III, 3, 5, 10).

The general point is that the structure of the cosmos shows that it is the result of design, and more specifically that it is designed for the sake of human beings.[3] Human beings have needs and the nonhuman world exists to satisfy these needs.

There are classic statements of anthropocentrism in Aristotle, in Aquinas—who with typical exhaustiveness offers some seven arguments in the *Summa Contra Gentiles* (Aquinas, 115–119)—in Immanuel Kant, and in the works of many other philosophers. Kant's position is typical: "Animals are not self-conscious and are there merely as a means to an end. That end is man. We can ask, 'Why do animals exist?' But to ask, 'Why does man exist?' is a meaningless question" (Kant, 1775–1780, 239; cf. Part I below). The otherness of animals, their very existence and their difference from human beings, is explained and understood in terms of this relation to human needs.

CRITIQUES OF ANTHROPOCENTRISM

In light of the history of environmental destruction and the heedless treatment of animals that is the Western legacy, especially since the advent of scientifically-based technology and industry, it is hardly surprising to see the appearance of fundamental critiques of the entire anthropocentric tradition. Some forms of this critique are focused on sentient animals, and these theories typically involve *extending* to animals the moral consideration that has traditionally been restricted to human beings. I discuss these approaches, animal rights and animal liberation theory (the latter two titles point to important philosophical differences), in Part I.

Other critics of anthropocentrism, instead of extending to the higher animals the moral consideration traditionally granted to human beings, develop ethical theories that are not dependent on traditional ethics. Such positions are *bio*centric—*life*-centered—in contrast to traditional *anthropo*centric—*human*-centered—positions. The first great pioneer of biocentric thought in twentieth-century Western philosophy was Albert Schweitzer, who developed his principle of "reverence for life" as a revolutionary answer to what he saw as the crisis of Western civilization. Schweitzer's ethical focus is not merely on human beings and animals, but on the world of life in its full breadth, which ultimately encompasses everything in our world (cf. Part II).

Biocentrism has deeper historical roots in American thought. In the mid-nineteenth century, American thinking about nature placed specific emphasis on the value of wildness and on the proper place of human beings in both nature and culture. In his essay "Walking" (1862), Henry David Thoreau protests against the destruction of wildness and insists that we will live better, more satisfying lives if we consider ourselves to be "an inhabitant, or a part and parcel of Nature" (Thoreau, 659), living "a sort of border life" between civil society and the wild (*Ibid*, 683). The wildness that Thoreau felt to be an essential aspect of authentic human selfhood required a new and different relationship to the wildness of nature, which Thoreau no longer regarded as something to be domesticated for the material benefit of domesticated human beings.

In *A Sand County Almanac* (1949), Aldo Leopold demands that we understand ourselves as "plain member and citizen" of the land community rather than as its "conqueror" (Leopold, 204). The task, as Leopold sees it, is to reframe our relationship to the natural world completely, to cease to see the natural world merely as raw material for whatever projects we think will enrich us, because we, like all living organisms, must live not only from the world of which we are a part, but also in the world. Above and beyond the instrumental value that nature can and does have for us, the biotic world has an inherent value in and of itself.

Such recognition requires a corresponding change in our understanding of ourselves. A new "You must change the way you live" confronts us, according to Thoreau and Leopold. This time the admonition does not come from the torso of Apollo—the idealized human body—that challenged the poet Rainer Maria Rilke to a higher, more truly poetic form of living (Rilke, 313/189). This time the challenge comes from a new recognition of the dignity of the natural world that our culture seems so bent on destroying. Such recognition changes our understanding of ourselves and of what it means to live a truly fulfilling and moral human life. Relating to wildness outside us, we cultivate the wildness within ourselves. The two sides condition one another reciprocally: a changed understanding of the meaning of the natural world brings about a changed understanding of ourselves, and vice versa.

CONTEMPORARY BIOCENTRIC EGALITARIANISM

Since the field of environmental ethics has been established as a recognized subdiscipline of philosophy over the past thirty years, the critique of traditional anthropocentrism has sharpened. Thoreau understood hunting and fishing to be expressions of human wildness and as conducive to an individual's development, even though he was drawn to the idea that vegetarianism is conducive to a higher spirituality.[4] Leopold was a passionate hunter. However, many contemporary philosophers have argued that the approval of hunting and fishing in Thoreau and Leopold is a residue of the anthropocentric attitude of domination each struggled so hard to overcome. The result of a radical critique of the presumption of human superiority has come in the form of *biocentric egalitarianism*. From this standpoint, hunting and fishing are viewed as expressions of the traditional assumption of human superiority over nature, and are therefore considered immoral. More generally, for biocentric egalitarians, any human actions that unnecessarily—in a very tough sense—harm nonhuman life are forbidden; it is morally impermissible for human beings to use nonhuman nature without some kind of overriding justification. This often takes the form of claims that we have a duty to protect and enhance the well-being of wild living things (Albert Schweitzer and Paul Taylor; cf. Parts II and III below) or that we are obligated to leave them alone to the greatest extent possible (Tom Regan, James Sterba, among others; cf. Part I below). Recognition of the inherent value of nonhuman life leads to recognition of an obligation not to harm such life.

THE FALLACY OF CONTEMPORARY BIOCENTRIC EGALITARIANISM

Such lines of thinking can seem very appealing as an antidote to the destructiveness, thoughtlessness, and general lack of concern for the natural world in our culture and its economy. But we must be careful about the way in which the idea of biocentric egalitarianism is developed. I argue that in a justified reaction against traditional anthropocentrism, much recent biocentric egalitarian thought has simply gone to the other extreme. Rejecting the idea that the natural world exists as a pure resource for human use, many argue that human beings are obligated, to the greatest extent possible, to refrain from using nature as a resource at all. The result is a curious mirror image[5] of anthropocentrism, as human beings are once again removed from nature. But this time the removal goes in the opposite direction from that taken by anthropocentrism. Instead of the removal making us lords of a nature that exists for our use, this new removal involves the moral obligation to take ourselves out of participation in natural processes to the greatest extent possible—out of respect for nature.

Of course this demand cannot be met completely, as many of these thinkers acknowledge. But the belief that this is the ideal toward which we should strive is widespread. Here are some examples:

1. Albert Schweitzer's principle of reverence for life requires that we "have the will to maintain our own life and every kind of existence that we can in any way influence, and to bring them to their highest value" (Schweitzer 1923, 278). Schweitzer recognizes that in this world, life lives at the expense of life, and

that necessity compels us to harm and destroy other living things so that we can live. But when we do so, Schweitzer insists, something evil occurs, and we incur guilt (cf. Part II below).

2. In their discussion of Arne Naess's philosophy of deep ecology, Bill Devall and George Sessions write, "Naess suggests that biocentric equality as an intuition is *true in principle, although* in the process of living, all species use each other as food, shelter, etc." (Devall and Sessions, 67; my emphasis). The clear suggestion here is that the recognition of a biocentric equality that would grant everything an equal right to its full self-realization, while a worthwhile ideal, conflicts with the reality that life lives from life. (It is not clear that Naess would agree with Devall and Sessions's interpretation.)
3. Paul Taylor, from the perspective of his principle of respect for nature, notes with regret that "Although we cannot avoid some disruption of the natural world when we pursue our cultural and individual values, [if we cultivate the attitude of respect for nature] we nevertheless constantly place constraints on ourselves so as to cause the least possible interference in natural ecosystems and their biota" (Taylor 1986, 310; cf. Part III below).
4. James Sterba writes, "As a moral agent, one's general obligation to all living beings is simply not to interfere with them" (Sterba 1998, 374).
5. Finally, even Aldo Leopold—for whom being a "plain member and citizen" of the land community is precisely to be a part, a *responsible* part, of the food chain, where good citizenship involves living in such a way that the flow of energy in the system is not impeded even as we take our biologically natural place within the system—slips into formulations that suggest that our status as *user* of nature as a resource is a regrettable fact. "A land ethic of course cannot prevent the alteration, management, and use of these 'resources,' but it does affirm their right to continued existence, and, at least in spots, their continued existence in a natural state" (Leopold, 204). Once again, the suggestion seems to be that the ideal would be total disengagement, but since this is impossible, we have to make do with a less radical disengagement.

I argue that something has gone deeply wrong here. The fundamental mistake common to all these thinkers is the assumption that there is a deep opposition between respect for the inherent value of living beings on the one hand and instrumental use of them on the other. If one makes this assumption, then it is not only natural but also inescapable to conclude that instrumental use of beings of inherent worth is *prima facie* a violation of that worth, *prima facie* a failure to respect.[6] For such a view, instrumental use may be justified in certain circumstances when the *prima facie* duty not to use is overridden by necessity, but the basic imperative against use remains as an ideal.

I argue—most formally in Part III, Chapter 2—that this assumption is based on the reduction of living things, human and nonhuman, to abstractions, all under the guise of assuming the moral point of view. When we think concretely, no living thing can be considered adequately in isolation from its place in a web of life in which everything lives by appropriating from the surrounding world the energy and nutrition it needs. The inherent worth of any living thing is inextricably bound up with the instrumental value of the energy and nutrition it appropriates from its environment, often enough from other living things; inherent value and instrumental value are intertwined with one another. To respect the inherent value of one being requires respect for its instrumental use of and to other beings. Put in Paul Taylor's terms, I show that there is no inherent worth (a being that has or pursues its own good) without instrumental value (that which is appropriated when such a being pursues its own good). This is not changed when that which is appropriated itself has inherent worth. Both instrumental value and instrumental use are thus constitutive moments of inherent value.

Human beings are no exception to this law of organic life and its value. The moral agent is not a disinterested spectator of the world of organic life, but an engaged participant in transfers of energy and nutrition—transfers of life itself—in the real world. Our being as organic forms of life requires that we participate in food chains; our being as moral agents requires that we ask *how* we can participate with respect for both those chains and the individuals that make them up—including ourselves. This is a dimension of human existence that has been neglected by most philosophers (Albert Schweitzer is a notable exception), since from an anthropocentric position this dimension raises few interesting issues. For traditional philosophy, the fact that human beings are organic forms of life is of importance chiefly because this is the cause of human mortality and suffering and a hindrance to the exercise of human rationality and freedom.

To call for the removal of human beings as participants in the natural order, even as an ideal, fails to respect both the inherent value of human beings as forms of organic life and the dignity of human appropriation. It is to sacrifice respect for our own organic being on the altar of moral principle. But I argue that moral obligation involves no such sacrifice, not even as an ideal. To recognize and respect the inherent worth of something does not commit one to the *prima facie* obligation to leave it alone, to refrain from making use of it or consuming it. The goal is rather to determine what constitutes the *morally respectful use and appropriation of* the natural world and of the beings of inherent worth with which we share that world. To consider how the lives of animals can be appropriated with respect, one must begin by recognizing the way in which the life of each living being is intertwined with the lives of other living beings. By the same token, one must begin with human beings who are integral parts of the organic web of life.

Two extremes must be avoided:

1. the unlimited appropriation and domination sanctioned by traditional anthropocentrism;
2. the ideal of zero human participation in and appropriation of the natural world demanded by biocentric egalitarians.

I discuss in some detail the views of several philosophers who have come to this kind of "egalitarian" conclusion: the animal rights theory of Tom Regan (Part I), Albert Schweitzer's ethics of reverence for life (Part II), and Paul Taylor's ethics of respect for nature (Part III).

PRIMARY APPROPRIATION: HUNTING AND FISHING AS A TEST CASE

For animal rights and current biocentric egalitarian approaches, nonsubsistence, "sport" hunting and fishing are the easy cases: given these ethical principles, any hunting that is not strictly subsistence hunting, for which there are no practical alternatives, is quickly seen as immoral. In contrast, an approach that emphasizes our respectful participation in food chains may come to very different conclusions about certain forms of hunting and fishing. For this reason, my concrete test case in discussing the theories of Regan, Schweitzer, and Taylor is contemporary nonsubsistence hunting.

By the same token, the contemporary practice of catch and release fishing hardly merits any discussion if a conventionally biocentric egalitarian concept of "respect for nature" is simply presupposed—it is just obviously wrong (cf., e.g., de Leeuw, who accepts this presupposition without any argument). For this reason, I take up the issue of catch and release fishing in Part V.

By taking as my test case what biocentric egalitarians consider to be easy cases, I pose a challenge, especially to Schweitzer and Taylor. I argue that hunting illustrates most clearly the positive value and meaning of human participation in natural processes in which life and death, self and other, are intertwined. The truly difficult and deadly serious issue is not hunting and fishing, but the relationship of a technologically advanced civilization to the natural world.

Hunting and catch and release fishing present something of a continuum. Beginning with subsistence hunting and fishing, in which the prey is required for sustenance, through "sport" hunting and fishing (including the eating of the prey), to catch and release fishing, in which the prey is released "unharmed" and which involves no biological appropriation, the continuum extends from necessary participation in the food chain, through ritualized and symbolic participation, to a practice that has no ties to the food chain. As Holmes Rolston III writes, "We move from eating—a primordial necessity—to play, seemingly trivial. Can we use these two activities to help us figure out who we are & where we are & what we ought to do?"[7]

I call hunting, killing, and eating a wild animal, even when it is not required for subsistence, the "primary appropriation" of animal life. In this series of actions our inescapable dependence on animal life is encountered in as direct a manner as possible. This dependence is inescapable, since even vegetarians and vegans engage in what I call "secondary appropriation," for all human actions have an unavoidable impact on animal life. Every time a field is cleared to grow organic vegetables, habitat for wild animals is destroyed. If there is an intrinsically moral dimension to our relations with wild animals, we cannot avoid the fact of universal human appropriation. I examine the moral questions raised by the practice of hunting in order to confront in as direct a manner as possible the moral dimensions of our dependence on nonhuman nature.

STRATEGIES FOR READING: ADVICE FOR NONPHILOSOPHERS

This book is a contribution to the field of environmental ethics, but it is written with the aim of reaching a much broader audience of interested nonspecialists. Specifically, reflective hunters and anglers should find the book to be interesting and rewarding. At the same time, some of the book contains fairly technical philosophical discussions, and a reader who does not want to tackle these sections right off the bat can easily avoid them and come back to them later.

A reader who is specifically interested in hunting and angling could go directly to Part IV, Chapter 2 and Part V. However, it would be more rewarding to take the following path through the book. Begin with Part I followed by the Appendix. These discussions are nontechnical, but give the reader a sense of what the more technical sections in Parts II and III are about. The next step would be to read the last section in Part II, Chapter 1, "Hunting and Reverence for Life," and the last section in Part II, Chapter 2, "Questions About Hunting." The reader will then be in a good position to read Parts IV and V.

AN OVERVIEW

In Parts I–III, a presentation or interpretation of the position under discussion is followed by a critique of that position. In each case, the critique is followed by a discussion of alternative approaches to understanding the moral status of hunting in the contemporary world, and these discussions of hunting make essential use of literature on hunting. My approach in these sections is broadly phenomenological, using hunting literature as a source of exemplary experiences and reflections on those experiences. My task is to analyze these experiences and reflections, looking for the essential structures and meanings that can be found in them or teased out of them. In Part IV I try to draw the various threads from the first three sections together. Finally, while catch and release fishing is not primary appropriation—and is often considered to be morally superior to "meat" fishing for that reason—this widespread practice raises moral issues, especially that of cruelty, which must be dealt with on their own terms. In Part V I use the results of Part IV to investigate these issues. The Appendix offers a critical study of Matthew Scully's important book *Dominion: The Power of Man, the Suffering of Animals, and the Call to Mercy*. It is included here because Scully's book is an excellent and important contribution to the political discussion of an important, if generally invisible, issue. My critique of Scully's philosophical underpinnings is a nontechnical version of my critiques of Schweitzer and Taylor in Parts II and III.

Acknowledgments

This book is the result of a series of encounters. The first encounter was with the writing of Irish philosopher A. A. Luce, who made me question a practice that is very close to my heart—catch and release fishing for trout. The second major encounter was with an essay by Barry Lopez, "Renegotiating the Contracts," which I first read while attending a National Endowment of the Humanities summer seminar on the American tradition of nature writing at Vassar College in 1997. The essay made me sit up and *listen.* Lopez, who offers his reader questions rather than answers, helped me reframe the questions Luce had confronted me with. It was my encounter with Lopez's essay that lead me to reencounter the philosophies of Tom Regan, Albert Schweitzer, and Paul Taylor with new questions.

I owe a great debt of gratitude to many writers on hunting: Stephen Bodio, George Bird Evans, Allen Morris Jones, Ted Kerasote, Buddy Levy, Thomas McGuane, Steven J. Meyers, Richard Nelson, Datus Proper, Mary Zeiss Stange, Jim Tantillo, Guy de la Valdéne, Richard Yatzeck, and many others.

As I wrote and rewrote, the work of several artists kept me company in various ways and were an important part of the process: Russell Chatham, Steven C. Daiber, David Friedman, Ladislav R. Hanka, Roxana Leask, Charles Lindsay, Jill Evans Petzall, and A. D. Tinkham. I am also indebted to Bob Dozier, Jerry Fisk, Koji Hara, Jeff Hollett, and Terry Primos for their consummate craftsmanship.

Several people read part or all of the manuscript at various stages, and their encouragement, feedback, and criticism have been a series of gifts: Larry May, Steven J. Meyers, Clare Palmer, Jill Evans Petzall,

Tom Regan, Holmes Rolston III, Mary Zeiss Stange, David Strong, Jim Tantillo, Paul Taylor, Red Watson, and Theodore Vitali.

As my thinking about hunting began to take shape, my mother, Maxilla Everett Evans, told me of visiting a swamp on the farm she grew up on with her brother, who was now the farmer. Dink Everett was a hunter, and when they found a winter flock of some 400 wood ducks, my mother was overcome by the question, "How can you kill such beautiful birds?" It is a question that has haunted my thought and my writing, just as it haunts a great deal of hunting literature.

Most important has been the constant love, encouragement, and questions from my wife, Jill Evans Petzall. She has not only helped me get clear about what I think, in part because she doesn't buy a lot of it and does not hesitate to challenge me on those issues. She has also contributed enormously to the clarity of my writing, so the reader owes her a debt of thanks as well.

Dedication

In February 1996 my father wrote his then weekly column in the local newspaper on "What's behind the decline of hunting in America?" In it he told of living through a cultural shift: "We grew up when hunting was a food-producing skill, and lived through its moving into a sport." Never the passionate and expert hunter and angler that his father had been, he slowly drifted away from first hunting and then fishing. As I began to think more deeply about these issues, I was happy to take up a dialogue with my father and, through him, with the grandfather who had died when I was still very young. I remember the last time my father and I went trout fishing, when he was in his mid-seventies. At dusk, he was watching me fish one last pool in the East Fork of the Pigeon River when a nice fish struck my dry fly. I missed the strike that even I had heard more than seen, and we stumbled back down the path that become too dark for his aging eyes when the rhododendron closed over our heads. I took his hand and talked him through the dark places.

In June 1999 my father wrote his column on "A trip through multi-infarct dementia and Alzheimer's disease." He used his own recent diagnosis of dementia to inform his readers about both dementia and Alzheimer's. He has lived the steady decline caused by his dementia with enormous grace.

This book is dedicated to Joseph Claude Evans, Sr.

Part I

R*E*S*P*E*C*T
Finding Out What It Means . . .

Chapter 1

The Challenge of Animal Rights and Animal Liberation Philosophy

INTRODUCTION

Some of the seminal work in what has come to be known as "animal rights" or "animal liberation" theory has involved extending to animals a moral status or a "respect" that has generally been accepted for human beings.[1] This extension is justified by the claim that refusing the extension is arbitrary and thus unreasonable. If human beings justifiably have a certain status, and there is no valid reason why only human beings should have this status, then we must extend this status until we have a valid reason to stop. Different thinkers stop at different places and for different reasons. This is what Holmes Rolston and before him Paul Shepard have criticized under the name "ethics by extension."

I will approach this topic by exploring some different senses of the term "respect," using hunting as a test case. After a very brief look at relevant aspects of animal liberation/rights theory I will focus on two uses of the term "respect" (there are of course many others), one concerning the respect due to human beings and one concerning the respect that some hunting cultures find due to wild animals. The first evokes a deep human experience of human beings while the second evokes a deep human experience of animals. My claim is that the term, as used by many animal rights/liberation thinkers, lacks any such relation to human experience and thus abstracts from both the reality of the animals and the reality of our experience of the animals. Ultimately, such thinking turns human beings themselves into abstractions. This raises the issue of the kind of respect properly owed to animals, specifically to wild animals.

Only against this background can we ask whether contemporary hunting in countries like the United States can be done with a proper respect for the animals hunted. This is not to say that I have found the high road to universal agreement. Far from it. But we—on each side of the issue—may come to understand both ourselves and our opponents better.

NOT AS "MEANS TO OUR ENDS"

Both Peter Singer and Tom Regan have worked out their positions in great detail, but for my purposes I can concentrate on what is of immediate relevance. For all their differences in philosophical principles, they share a line of thinking with many other animal rights, animal liberation, or animal welfare theorists, and I want to put this before us for discussion. Singer and Regan appear here only as examples of a much more widespread tendency, one which can be expressed in many different ways. While I will be focusing on discussions of what can be called "respect," appearance of the word itself is not essential.

Peter Singer's approach is utilitarian, but he went beyond mainstream utilitarian thinking in picking up and developing a neglected aspect of the work of Jeremy Bentham, who wrote that the question "is not, Can [animals] reason? Nor, Can they talk? But, Can they suffer?" (Bentham, ch. 18, sec. 1). Singer argued that since animals can suffer, they undeniably have an interest in the outcome of many if not most human actions, so their interests should be taken into account in moral reasoning. Singer thus attacks what has become known as "speciesism," the view that only the members of one species, the human species, deserve direct moral consideration. If we are to avoid speciesism (just as morality requires that we avoid racism and sexism), the suffering of any sentient being must be given equal weight to the like suffering of any other sentient being. Treating nonhumans as "utilities," as "means to our ends," is therefore immoral (cf. Singer 1998, 100, 101).

Singer was soon challenged from within the ranks of those concerned with reforming our relations with animals. While Tom Regan was sympathetic with many of Singer's practical conclusions, he disagreed sharply with Singer's philosophical principle, with his utilitarianism. Regan argues that Singer's utilitarian approach fails to value individuals properly, be they human or animal, since utilitarianism's aggregative approach does not recognize individual rights. For Regan, any being that has "inherent value" has rights. So which beings have inherent value? Regan, like Singer, argues that there is no reasonable way to limit recognition of inherent value to human beings alone. He argues that the only nonarbitrary position involves recognizing that any being that is "the subject-of-a-life" has interests that are to be respected (Regan 1983, 243–248), and that justice requires that "we are to treat those individuals who have inherent value in ways that respect their inherent value" (*Ibid*, 248–250). He joins Singer in arguing that "individuals who have inherent value must never be treated *merely as means*" (*Ibid,* 249), though Regan charges that any aggregative approach, such as Singer's, will in fact violate this principle.[2]

Singer's position has come to be known as "animal liberation" in contrast to Regan's "animal rights" position. But in spite of their differences, Singer and Regan have importantly similar approaches with regard to broader environmental issues. Singer urges a "hands-off" approach to animal life, and to wild animals in particular: ecosystems should not be "managed," even with the proclaimed goal of benefiting wildlife. Regan goes even farther than Singer, arguing that any attempt to protect species and/or ecosystems at the expense of individual animals is "*environmental fascism*" (*Ibid*, 362). Thus, for Regan, only the individual members of a species are worthy of direct moral concern, not the species itself, and the individual members of an endangered species, as individuals, are no more worthy of moral concern than are any other individual animals.

"RESPECT" IN ANIMAL LIBERATION/RIGHTS THEORY

A theme is emerging from this brief look at animal rights theory. Though the details, the language, and to some extent the practical upshots vary, Regan and many other theorists claim something like the following:

> *If something has inherent value, it is to be treated with* ***respect****, where respect requires that we not treat that thing as a mere means to our ends.*[3]

This general position has been given a succinct formulation in a recent essay critical of catch and release fishing. A. Dionys de Leeuw, a professional biologist specializing in sport fisheries management, sums up more than twenty years of thinking about animal rights when he gives the following definition of "respect": "behavior with regard to an interest that shows consideration for the holder of the interest and avoids degradation of it, negative interference with it, or interruption of it" (de Leeuw, 379).[4] Given this definition, which de Leeuw simply accepts without any discussion of the background in his discussion of the ethics of angling (this is an indication of just how deeply ingrained this approach has become), respect for an animal requires that (*ceteris paribus*) we avoid interfering with that animal as it pursues its interests.

"Respect" is of course a technical term in the moral theory developed by Immanuel Kant in the late-eighteenth century, and I think that the ghost of Kant haunts much recent thinking about animals. Kant requires that we treat "persons" with *respect*, which forbids us to use them merely as means to our own ends. Formulations such as "means to our ends" (Singer) or "merely as a means" (Regan) derive from Kant, and the reference is often made explicit (e.g., by Regan). But to demand that animals be treated with "respect" also requires that one reject some of Kant's claims. After all, it was Kant who stated that we have no direct duties to animals, since he held that animals are properly only means to an end for human beings. "Our duties towards animals are merely indirect duties towards humanity" (Kant 1980, 239). Since animals are not "persons" in Kant's sense, to require that we treat them with the respect proper to persons simply makes no sense.

Animal rights thinkers will quickly reply that this is no reason not to extend to animals a basic moral considerability, and I have no quarrel with them on this point. But I do want to suggest that if that recognition of moral considerability occurs by extending to animals the respect that Kant argued is due to persons, the results are counterintuitive. Curiously enough, the result of the work of Singer and Regan (and many who have followed in their footsteps) is that for the animal rights/liberation positions, the demands of respect are actually *more* rigorous (as opposed to being equally rigorous, which seemed to be the aim) regarding animals than regarding human beings. Kant allows, indeed insists, that persons may compete with one another as they pursue their individual visions of happiness. In so doing one's actions may have a detrimental effect on, and in that sense may "negatively interfere" with, the projects of another human being as long as one does not treat that person "*merely* as a means." In other words, our interactions with other persons are limited by rules—moral laws and positive laws—but these are rules governing specifically human interactions. I may outcompete you in the economic arena, but I am not allowed to steal from you. This scenario is not possible when it comes to our relations with animals, since we cannot converse with them. So in the place of the rich rule-governed interaction of human beings, we have the "keep your distance" of a hands-off policy when it comes to animals.[5] "Respect" in this sense requires nonparticipation (always under a *ceteris paribus* qualification). But I want to suggest that something has been lost here, with pernicious consequences. We can see these pernicious results more clearly by looking first at the precise role of "respect" in Kantian moral theory, and then contrasting it with a form of respect for animals that involves complex interrelationships between humans and animals—the very opposite of a "hands off" approach.

KANT

I have suggested that the strategy of extending the Kantian principle of respect to animals loses contact with the palpable reality of animals in human experience, thereby reducing them to abstractions. We can perhaps find a clue to what has gone wrong by looking at the Kantian principle of respect in the context in which it is developed in order to see why it is appropriate *to that context*. This may allow us to see why it is inappropriate when extended to a different context. The brief interpretation of Kant's approach to ethics that follows differs from many currently accepted approaches to Kant. I think, however, that it is closer to what Kant is actually doing in his ethics than what is given in some of these other interpretations. More importantly in this context, since it does not lend itself to ethics by extension, it gives food for thought, even if one does not accept the Kantian position itself.

For Kant, what is distinctive about human beings is that they are rational: only rational beings have "the capacity of acting according to the *conception* of laws (i.e., according to principles)" (Kant 1990, 29).[6] By the same token, only a rational being can respond to being questioned—by oneself or by another—by giving the *reason why* he or she acted in a certain way, by *justifying* the action. The reason offered can be satisfying in one of two different ways. First, we may see that the action could reasonably be expected to produce some goal that the person acting actually has, ultimately by contributing to the person's happiness. Whether another person shares that goal—whether it would be in any sense satisfying to that person—is irrelevant. But there is a second way of considering the action. We may ask why it is *morally permissible* for a person to act in a certain manner even if doing so would undoubtedly contribute to that person's happiness. Here Kant in effect asks what constitutes a justification that can claim to satisfy any possible interlocutor, arguing that any morally permissible action must satisfy one condition: any action that is morally permissible for you in a certain situation must be morally permissible for me, indeed for any moral agent, in a relevantly similar situation. In short, moral reasons or justifications must be *universal*, applicable equally to you and me and to every moral agent. The rationally unacceptable alternative is that we ultimately have to say that it is morally permissible for me, e.g., to steal a car I desire, because I am me, while it is not morally permissible for you to steal the car (perhaps from me!) because you unfortunately are not me. This is not something that I or any one else can rationally accept as an adequate justification.

This consideration immediately leads to Kant's initial formulation of the moral law, the categorical imperative: "Act only according to that maxim by which you can at the same time will that it should become a universal law" (*Ibid*, 38). This means that I must in principle be able to explain to any other person why my action is morally permissible: any action which results from following a maxim which can be willed to become a universal law is permissible. This is something that I can demonstrate, and as a rational being, I am responsible to other rational beings in this sense: I must be able to justify my actions.[7]

For Kant, this shows that those beings to whom I must justify my actions—to whom I am responsible in the sense specified—have a different status than that of beings with whom I cannot jointly consider moral justification. Rational beings thus have "absolute worth" in a very specific sense. To have absolute worth is to be a being to whom I am responsible, again in the specified sense, because that being is rational (a "person"), because that person can ask whether an action someone wants to do is morally acceptable. Each rational being is an "end in itself," having a right to demand acceptable moral justifications. Since rational beings have absolute worth, they are deserving of our "*respect*" in a precise technical sense: the moral law requires that one "Act so that you treat humanity, whether in your own person or in that of another, always as an end and never as a means only" (*Ibid*, 46).

Two things about the Kantian position as I have presented it are important when we think about the moral status of animals. First, it does not make sense to extend to animals the specific respect his theory shows to be due to persons.[8] I cannot be responsible to, in the sense of being responsive to the legitimate demand for justification from, a creature that cannot itself act according to principles and therefore cannot engage in deliberation about justification. Kant's principle of respect is rooted in my concrete, deliberative relations with other *rational* beings, beings with whom I can discuss moral questions, and it is ultimately their capacity to be rational that I, if I am rational, must respect. (Put more strongly: *if* I am truly rational then I *will* respect them.) Any attempt to extend *this* kind of respect to animals cuts the principle of respect off from such situations of moral discussion and justification without replacing them with other forms of interaction. This seems to confront us with an either/or concerning animals that are not persons. Either we follow Kant in considering them as mere things to be manipulated for our own benefit, or, if we do want to acknowledge their moral standing, the only permissible thing seems to be to extend a truncated sense of the respect due to persons to animals, with the result that we must leave the animals alone, literally to have nothing to do with them.[9] If I must be responsible to an animal that is not itself responsive to reasoned discussion, all I can do is stay away, since there is no way of determining what an acceptable interaction would be. It is as if moral principle demands that we not be allowed to inhabit the same world with the animals, since we do not live in the same moral world.

But, second, it is important to note that Kant's ethical theory, as I have presented it, is not necessarily committed to a traditional hierarchy that gives an absolute privilege to human beings. Nothing in Kant's position as presented here implies that we have no direct duties to animals, that there is not a specific moral consideration due to animals. The only thing we can conclude is that the moral status of animals will have to be based on something other than the status of persons. The fact that Kant himself says that we have no direct duties to animals does not mean that his basic ethical theory implies it. In other words, Tom Regan is wrong when he states that Kant's claim that we have only indirect duties to animals is "a direct consequence of his moral theory" (Regan 1983, 175)—wrong, that is, if we consider the core I have presented to be the essence of this moral theory. (That it follows from other things Kant is committed to I certainly acknowledge.) Regan acknowledges that Kant's position (and that would include Kant's position as I interpret it, I think), is not a form of contractarianism, and thus not objectionable on those grounds. What I am arguing is that the arbitrariness Regan finds in Kant's ethics does not affect its theoretical core.

This point becomes crucially important if we agree with Singer and Regan that exclusion of animals from moral consideration is arbitrary—and I have no argument with them on this point. For Kant, animals are not eligible for the kind of respect due to persons, and Kant concludes that they are therefore mere things to be used by human beings as we see fit, *mere* means to our ends. If one rejects this conclusion, *and if one accepts the claim that this is an exhaustive either/or—respect OR use as mere means*—the only way out seems to be simply to extend his demand for respect for persons to animals. This is exactly what we saw Singer and, particularly, Regan do when they use phrases like "means to our end" and "simply as a means."[10] But this ignores the possibility that the *either/or* is not exhaustive.

Thus, we have to consider the possibility that there may be *other* forms of respect which are appropriate to our relations with animals. While I think that Kant's critics are right in rejecting his claim that we have no direct duties to animals, that they are just instruments for us to use in any way we see fit, I think that they move too quickly to the program of ethics by extension. What, then, might a concrete, nonarbitrary form of respect for animals look like?

RESPECT AND MYTHIC RECIPROCITY

In an essay entitled "Renegotiating the Contracts," Barry Lopez argues that our relationships with wild animals have changed in the modern world. As he puts it, " . . . our relationships with wild animals were once contractual—principled agreements, established and maintained in a spirit of reciprocity and mythic in their pervasiveness. . . . these agreements derived from a sense of mutual obligation and courtesy" (Lopez 1991, 381). Lopez is not just using a weak or misleading metaphor. His talk of "contracts" is based on the idea that "We once thought of animals as not only sentient but as congruent with ourselves in a world beyond the world we can see, one structured by myth and moral obligation, and activated by spiritual power" (*Ibid*, 382). Many hunting cultures speak of a time when humans and animals spoke a common language, and in such cultures stories about encounters with animals are crucial to learning the art of living a successful life.

In *Arctic Dreams* Lopez writes, "The evidence is good that among all northern aboriginal hunting peoples, the hunter saw himself bound up in a sacred relationship with the larger animals he hunted. The relationship was full of responsibilities—to the animals, to himself, and to his family" (Lopez 1987, 199). Later Lopez writes, "Hunting in my experience—and by hunting I simply mean being out on the land—is a state of mind. All of one's faculties are brought to bear in an effort to become fully incorporated into the landscape . . . To hunt means to have the land around you like clothing. To engage in a wordless dialogue with it, one so absorbing that you cease to talk with your human companions. It means to release yourself from rational images of what something 'means' and to be concerned only that it 'is' (*Ibid*, 199–200). It is this kind of intimacy that Lopez thinks we have lost. For Lopez, we have indeed come to view animals "merely as means," as commerce views animals either as mere commodities (e.g., chickens in a chicken factory) or as hindrances to commerce (e.g., spotted owls to loggers and wolves to ranchers). And he considers this most fundamentally "a failure of imagination. We have largely lost our understanding of where in an adult life to fit the awe and mystery that animals excite" (Lopez 1991, 384), an awe Lopez finds alive in the hunting cultures in which he has lived. And against those who would say that this is to be welcomed as progress beyond the stage at which human life is necessarily dependent on the exploitation of wild animals, Lopez argues that "to set aside our relationships with wild animals as inconsequential is to undermine our regard for the other sex, other cultures, other universes" (*Ibid*, 383). In short, our relationship with wild animals is one of those nodal points at which our relationship with otherness as such is formed. (For an extended argument for the importance of wild animals in human experience see Paul Shepard, *The Others: How Animals Made Us Human*.)[11]

But what is at stake here runs even deeper. Lopez writes:

> *No culture has yet solved the dilemma each has faced with the growth of a conscious mind: how to live a moral and compassionate existence when one is fully aware of the*

> *blood, the horror inherent in all life, when one finds darkness not only in one's own culture but within oneself. If there is a stage at which an individual life becomes truly adult, it must be when one grasps the irony in its unfolding and accepts responsibility for a life lived in the midst of such paradox. One must live in the middle of contradiction because if all contradiction were eliminated at once life would collapse. There are simply no answers to some of the great pressing questions. You continue to live them out, making your life a worthy expression of a leaning into the light* (Lopez 1987, 413).

The parallels and differences between Lopez and animal rights advocates can be conveniently and dramatically pointed out by means of a comparison with Peter Singer. Singer himself is very upfront in admitting that he is neither particularly fond of nor even interested in animals. His concern is focused on a matter of principle, on "ending oppression and exploitation wherever they occur, and in seeing that the basic moral principle of equal consideration of interests is not arbitrarily restricted to members of our own species" (Singer 1990, ii). But while Lopez is critical of our overwhelmingly commercial relationship with animals, and thus joins Singer and Regan in opposing factory farming and animal research, he can have little sympathy with the animal liberation/rights positions we have discussed, since they require that our loss of contact with wild animals, which Lopez diagnoses and mourns, be intensified as a matter of principle. The lack of genuine contact with animals found on the factory farm finds an ironic counterpart in the lack of contact of the "hands-off" position. And while Lopez himself is clearly uncomfortable with killing, I think he would argue that this is not the central issue in our disturbed relationship with wild animals. More important is the fact that, as he puts it,

> *[W]e [as opposed to Eskimo culture] have irrevocably separated ourselves from the world that animals occupy. We have turned all animals and elements of the natural world into objects. We manipulate them to serve the complicated ends of our destiny. Eskimos do not grasp this separation easily, and have difficulty imagining themselves entirely removed from the world of animals* (Lopez 1987, 200).

So Singer's lack of interest in animals as such is for Lopez precisely part of the problem, and the alternative for him is anything but a "hands-off" policy.

Lopez would not, I think, admit that he sanctions treating animals merely as means, since that would be precisely to deny the spiritual dimension he thinks we need to recover. Against Kant he would hold that it is wrong to divide the world into *persons* (rational beings, moral agents) worthy of respect and *things* (including animals) which merely have a price, since their value consists in their usefulness to humans. There is an enormous middle ground of beings worthy of an appropriate respect that is, however, different in kind from the respect appropriate to a moral agent. These beings are to be respected for what they are, and what they are puts them into a complex set of relations with the rest of the world, and thus with us. Among these relations we find those of hunter and hunted, the eater and the eaten.

From this point of view, the mistake in animal rights theory is to attempt to correct Kant by simply extending the rights—the "respect"—due to persons to the animals Kant neglected, as if that were the only sense in which we can recognize their moral status. The hunting cultures Lopez has studied recognize both our kinship with the animals we hunt as well as the differences between us. From this perspective, that animal rights and animal liberation positions erase the differences between animals and humans in an overreaction to Kant's claim that we have no direct duties to animals, that they are not themselves worthy of moral consideration. Their mistake is, as it were, a mirror image of Kant's. And it is only when we reduce the rich reality of the animals, who exist in a complex natural environment that includes complex human beings, to an abstract bundle of interests or to an equally abstract "experiencing subject of a life," that this simplistic extension of rights seems plausible.

RESPECT AS CONGRUENCE

But how does Lopez's account help us think about issues like contemporary hunting? The "contractual" relationships with wild animals he describes are part of a myth-pervaded experience of the world that is not available to those of us not born into it. Even someone like anthropologist Richard Nelson, who lived with and studied Eskimo and Koyukon hunting people for long periods of time (cf.

Nelson, 1969, 1983), recognizes that he does not live in their world (Nelson 1997, 286). (And Nelson did not just study these people as a scientist—the elders whose words repeatedly come back to him are clearly his mentors.) We may accept Lopez's critique of our commercial relationship with animals, but how can he help us regain a positive relationship with wild animals?

What is ultimately at stake is our relationship to the life of which we are part. An integral aspect of this life is something that at first seems to be its antithesis, death. Life itself, when viewed as a process, is inextricably bound up with death, as one being dies in order that another may live, and not even the most radical veganism can change this fact. Failure to see *and affirm* this reciprocal relationship of life and death, of self and organic other, leads to the kind of dissociation from life in the name of morality we have seen in Singer and Regan, a dissociation which reaches its paradoxical extreme in Cleveland Amory's famous statement that if he ruled the world, not only would hunting be prohibited, "Prey will be separated from predator, and there will be no overpopulation or starvation because all will be controlled by sterilization or implant" (Amory, 136; cf. also Sapontzis). Note that in such an approach, intervention and control are pervasive. But more importantly, the fact that such isolation would rob wild animals of their wildness, effectively putting a stop to natural evolution and immediately leading to a genetic decline in animal populations, is enough to show that the fundamental attitude at work here is one of enmity or at least opposition to the very process of life itself. As J. Baird Callicott has argued, "the value commitments of the humane movement seem at bottom to betray a world-denying or rather a life-loathing philosophy. The natural world as actually constituted is one in which one being lives at the expense of others" (Callicott 1992a, 55). This is an ironic result: what started out as "respect" for sentient beings ends up rejecting on moral grounds the conditions on which the cycle of life, which includes moral agents, is possible in the first place.

For Lopez, respect for wild animals is not the kind of respect for animals or for animals' interests that leads to a hands-off ethic. Nor is respect rooted simply in the sentience of animals. Rather, respect has to do with the way we as a species fit into the broader world in which we are inextricably interwoven. "If we could establish an atmosphere of respect in our relationships, simple awe for the complexities of animals' lives, I think we would feel revived as a species" (Lopez 1991, 386). And this respect and awe are a crucial part of what it means to live a truly successful life—successful in human terms which are not dissociated from the world of which we are a part.

> *The aspiration of aboriginal people throughout the world has been to achieve a congruent relationship with the land, to fit well in it. To achieve occasionally a state of high harmony or reverberation. To dream of this transcendent congruency included the evolution of a hunting and gathering relationship with the earth, in which a mutual regard was understood to prevail* . . . (Lopez 1987, 297).

This is not the congruency we feel between Woody Allen and New York City, the congruency between human beings and an increasingly artifactual environment, where food, be it meat or vegetable, comes from the store in the mall. In our urban society, this goal of transcendent congruency does not even make sense to many people, since in their experience it falls between the stools of the artificial environment we produce and live in and the transcendence of God. But it is also felt by many—perhaps as nostalgia, sometimes as one of the overriding goals of living one's life. And there are many kinds of activities in which we can have a heightened sense of such congruency. Hunting is only one such possibility. (*Being hunted*—by an animal—is another, terrifying, such possibility.)

But what about the "mutual regard" Lopez speaks of? Not only do we not live in the kind of myth-pervaded world of indigenous hunting cultures, only a very few of us can claim that our hunting is anything close to the subsistence hunting of those cultures. Lopez is surely correct when he writes, "The hunting contracts of our ancestors are no longer appropriate . . ." (Lopez 1991, 387). So how are we to think about hunting—the forms of hunting possible for us? Is it even possible to find a proper form of respect for animals in contemporary hunting? Is a new hunting contract possible?

I do not want to pretend to answer this question here, but an initial survey of some of the terrain to be explored may be in order. In what follows I will use a few texts from hunting (and antihunting) literature in an attempt to mark out some of the essential possibilities we encounter here.

THINKING ABOUT HUNTING

To think about hunting in our culture today is to think about what is called sport hunting. The term itself seems to demonstrate that the problem of a new contract is insoluble, that all we are left with is idle amusement . . . at the expense of the lives of animals. Some hunters seem to admit as much. For example, E. Donnell Thomas seems to think that the fact that hunting is fun is all the justification it needs, and he reduces all objections to hunting to a puritanical rejection of any enjoyment of life. "In short, we have become a society of secular Puritans, reading evil into just about anything that anyone could possibly do just for the hell of it" (Thomas, 6).[12]

Vance Bourjaily takes a similar line, and his argument is unintentionally telling:

> *Now there is no honest defense of any pleasure except to say: I do it because I enjoy it. When criticized we are likely to take peripheral benefits (exercise, identification with tradition, relaxing of tensions) and try to make them stand up as central justification. This seems to me a mistake. All we ought really to say to those spoilers who would suppress pleasures they do not share is this: disapprove of me as you will but to try to give your disapproval the force of law is a crime against freedom* (Bourjaily, 76–77).

The invocation of liberal values of individual freedom is impressive, of course, but perhaps this is to move too quickly, to miss the complexity of the phenomenon, the problems it engenders, and the justifications which amount to more than "I like to do it."

A more neutral definition has the virtue of leaving open a field for inquiry. Brian Luke defines "sport hunting" as "hunting done for its own sake, in contrast to subsistence hunting (done as a means of survival) and market hunting (done to sell parts of the animals' bodies)" (Luke 1997, 25). There is room for complexity in the "for its own sake," leaving open the possibility of more complex interpretations of the "sport" in sport hunting. Stephen Bodio writes:

> *... consider those often-misunderstood concepts "hunting" and "play." Sport hunting is not (despite an animal-rights brochure I read recently that blandly asserted it was, "of course," sublimated sexual sadism) some sort of aggression against creation. It is a series of rituals that have grown up around the most basic activities: acquiring food—capturing energy to keep us alive. Some of the rituals have come about because of their beauty, grace, and difficulty; others (like the German custom of giving the fallen animal a sprig of its favorite food), because of the sadness and mystery that accompany taking a life.*
>
> *Hunters who hunt out of physical need still appreciate these rituals; ones who do so out of "play" or out of a civilized desire to personally touch the roots of the flow of energy may elevate the ritual to the end result. The finest kinds of hunting—fly fishing, falconry, upland shooting with pointing dogs—are, and should be, elaborate ways of playing with your food and with the universe, ways that also give you windows into the lives of things as alien as insects (in fly fishing) or into the minds of canine and avian partners. Ideally, you leave the human behind for a few moments and become predator, prey, nonhuman ally* (Bodio 1997, 230).

The depth of the shift from Thomas's "fun" and Bourjaily's "my pleasure" to Bodio's almost cosmic concept of "play" can hardly be overemphasized. More importantly, continuity between some forms of contemporary "sport" hunting and the myth-pervaded, "respectful" hunting of some indigenous peoples becomes visible here.

But the notion of "playing" with the very lives of other creatures will be repugnant to many who have no trouble embracing the fundamental intertwining of life and death. And there is indeed a one-sided emphasis on a shallow meaning of "sport" in much of contemporary hunting (and hunting literature). Here I think that we must listen carefully to Joseph Wood Krutch when he writes:

> *Killing "for sport" is the perfect type of that pure evil for which metaphysicians have sometimes sought. Most wicked deeds are done because the doer proposes some good to himself ... The killer for sport has no such comprehensible motive. He prefers death to life, darkness to light. He gets nothing except the satisfaction of saying, "Something which wanted to live is dead"* (Krutch, 148).

This is powerful stuff, and we should listen carefully. But we must also retain our ability to make essential distinctions. When does our play become nihilistic? Here we are on an edge, and many paths will part at this point. What is perhaps most important is getting clear on why, so I want to approach the edge slowly.

Compare Krutch's words to the following passage from Stephen Bodio's memoir, *Querencia*:

> *One evening we were relaxing with drinks after dinner at the house of some good, very civilized friends, watching a PBS nature film. The usual cheetah began the usual slow-motion chase after the usual gazelle. The music swelled to a crescendo then stopped dead as the action blurred into real-life speed, dust, and stillness. Betsy and I raised our glasses and clinked them. Our hostess had left the room and her husband looked at us, puzzled. "You know," he said, "You're the only people I've ever seen who cheer the bad guys in the animal shows"* (Bodio 1990, 28).

Is this the expression of the nihilism Krutch finds intrinsic to sport hunting? Are the satisfactions of hunting essentially different from the pleasure taken in the beauty of the cheetah's kill and the deep love of the life—which includes death—of which both cheetah and gazelle are a part? This is a serious question, and Louis Owens, for one, rejects turning "the uncomplicated reality of this thing [a coyote killing a fawn as the doe tries to protect it] into a dreadful aesthetic" (Owens, 184). And yet is there not the possibility of watching such a scene unfold in awe, perhaps tinged with horror, but not with disgust? After reflecting on the experience of discovering that a mountain lion had been stalking him as he fished for trout, Owens writes, "It is in the end the awful beauty of the dance of deer and coyote that I remember best from that summer day" (*Ibid*, 187). I say that we should listen carefully to Krutch not because I agree with his general claim, but because our attention perhaps becomes properly focused on human hunting when we experience Krutch's words as a slap in the face. The kind of nihilism Krutch is describing is a permanent and constant possibility for hunting in our culture—a pure pleasure in domination and destruction. But this should not keep us from asking whether there are other possibilities here as well, since much antihunting thought (like the brochure Bodio cites) insists that there are no essentially different possibilities. To shift the focus back to human hunting, compare Krutch's lines with the following two passages written by Thomas McGuane. The first passage is a reflection on the occasion of killing a pronghorn.

> *Nobody who loves to hunt feels absolutely hunky-dory when the quarry goes down. The remorse spins out almost before anything and the balancing act ends on one declination or another. I decided that unless I become a vegetarian, I'll get my meat by hunting for it . . .*
>
> *A world in which a sacramental portion of food can be taken in an old way—hunting, fishing, farming, and gathering—has as much to do with societal sanity as a day's work for a day's pay* (McGuane 1982, 230, 236).

Here the whiff of nihilism—the pure preference for death over life—that filled Krutch's nostrils dissipates, is replaced by a deep and (to some) satisfying odor of life itself, which includes death. This is precisely the odor Cleveland Amory could not stand. As James Swan notes, "There is a lightness to the word *sport* that I think does a disservice to hunting. Like his predecessors, the modern hunter hunts for meaning, to express himself as a member of the human race" (Swan, 144; cf. also Stange).

Rather than focusing on concepts such as "pleasure," "fun," or "recreation," I think that it is more productive to begin with the concept of "satisfaction" as it appears in the following passage from an interview with Richard Nelson. In speaking of his experience hunting while living with Inupiaq Eskimo, Nelson says the following:

> *Another thing about hunting that struck me was a personal thing. For the first time in my life, I found myself engaged in the entire process of keeping myself alive, and it was a tremendous breakthrough in my understanding of where my life comes from. I remember wondering why this hunting life was so satisfying. Part of it was that I was involved with the whole process of keeping myself alive, from the often laborious and lengthy process of finding an animal to killing it, taking it apart, and then learning how it becomes food. I had never done any of that before. Food had always come out of*

> *the store. The deep sense of satisfaction I discovered in that process has never changed* (Nelson 1994, 82).

This need not be anthropocentric in any narrow sense (i.e., in the sense of using animals as "mere resources"); something of the "transcendent congruency" Lopez speaks of becomes apparent here. In other words, hunting just might be a proper way for a moral agent that understands him or herself as being part of the world to relate to wild animals *as such*. This suggests that, *contra* animal rights/liberationist theory and much of the thinking that emphasizes "respect for nature" and biocentric egalitarianism (cf. Part III), a proper respect for wild animals and for ourselves *requires* that we relate to them—in a manner that is respectful of their wildness, of course. But this is a long way from the abstract and isolationist "respect" for animals that dominates much of the literature. Indeed, from this perspective one way of respecting precisely their (and our) wildness might be to hunt them.[13]

But the possibility Krutch points to remains. Some hunters do hunt simply in order to kill (cf. Fontova, 54–55), but most hunters who reflect on their hunting deny this. The classic statement is found in Ortega y Gasset's *Meditations on Hunting*: "one does not hunt in order to kill; on the contrary, one kills in order to have hunted" (Ortega 105; cf. Part IV below). This leaves the question of why one hunts to begin with, and the answers that have been given to this question are many and varied (cf. Wood, 16–38). As the quote above from McGuane shows, hunting, for many reflective hunters, is an activity that takes place on the edge, indeed on many edges.[14] And this is one of them: either hunting is or can be an expression of our sense of ourselves as a respectful and responsible (and playful!) part of the world we inhabit, or it should be consigned to an earlier stage of culture, one which, for better or worse, we have left behind. In other words, the issue is not just whether hunting can be ethically justified. A morally sound relationship to wild animals is threatened not only by development that destroys habitat, but also by "preserving" habitat in the form of closed game ranches. So the question remains whether hunters can rise to the challenge. Hunting, far from being intrinsically an expression of human domination, makes demands on us, and we have to ask whether we today can still rise to these demands. Without responding to those demands, without an appropriate respect, no "new contract" is possible.

CONCLUSION

It is important to be clear as to what I have and what I have not attempted to do in this discussion. I have not tried to give a refutation of animal rights and animal liberation thought. I have also not tried to give a refutation of the general program of ethics by extension. What I have tried to do is to sketch an alternative way of approaching the issue of the human relationship to animals, particularly wild animals. The first goal in developing this alternative is to remove the sense of inevitability that is an important aspect of the rhetoric and argument of ethics by extension. If one takes this alternative seriously, extensionist arguments will, I think, be less compelling. This will not shake the convinced defender of animal rights or animal liberation, but it may give someone considering these positions pause. What is ultimately at issue is who we are. Answering this question requires not only that we inquire into the ways in which our sense of ourselves determines our sense of the world we relate to in all of our actions, but also the ways in which our sense of the world of which we are a part determines our sense of who we are.

If one does follow my alternative, however tentatively, one finds oneself in a different world of both experience and thought about the relationship between human beings and the rest of the animate world. But one could equally argue that the idea of a "new contract" that would be different from the recognition of either animal rights or equal consideration of interests, is romantic nonsense. I shall argue that just the opposite is the case. To do this, in Parts II and III I discuss two approaches, Alert Schweitzer's principle of reverence for life and Paul Taylor's ethics of respect for nature, that are not extensionist in their arguments, but which arrive at positions that are in many ways similar to those of animal rights and animal liberation thought. My goal in these parts will be to develop internal critiques demonstrating that neither position is philosophically and morally adequate. In short, my aim will be refutation in a philosophically rigorous sense. In addition, I shall argue that the *ways* in which they turn out to be in need of correction point in very different directions than animal rights or animal liberation theory. If I succeed in this project, I will have laid a better foundation for developing a concept of "respect" for animals and for nature in general, one that is neither an extension of nor modeled on the Kantian concept of "respect for persons."

Part II

Albert Schweitzer
The Principle of Reverence for Life

Chapter 2

Albert Schweitzer's Philosophy of Reverence for Life

INTRODUCTION

"Ethics is responsibility without limit towards all that lives" (Schweitzer 1923, 311).[1] For Albert Schweitzer, reverence for life, *Ehrfurcht vor dem Leben*, *veneratio vitae*, is the fundamental principle of ethics, and he became famous for, among other things, his inclusion of *all* forms of life under this fundamental principle, to the point that when he saw a worm on the hard path, he would stop and put it back into the kind of soil it required.

The principle of reverence for life came to Schweitzer very suddenly in 1915 as he wrestled with the problem of how "a culture which would possess a greater ethical depth and energy than our own" might be developed (Schweitzer 1963, 179). He systematically worked out this approach to ethics in the early 1920s. But his deep conviction that it is wrong to harm animals was rooted in experiences from his early youth. In his *Memoirs of Childhood and Youth*, he writes,

> *As far back as I can remember I was saddened by the amount of misery I saw in the world around me . . . A deep impression was made on me by something that happened during my seventh or eighth year. Henry Braesch and I had used strips of India rubber to make slingshots, with which we could shoot small stones. It was spring and the end of Lent, when one morning Henry said to me, "Come on. Let's go on to the* Rebberg *and shoot some birds." To me this was a terrible proposal, but I did not venture to refuse for fear he would laugh at me. We got close to a tree which was still without any leaves, and on which birds were singing beautifully to greet the*

> *morning, without showing the least fear of us. Then stooping like a Red Indian hunter, my companion put a ball in the leather of his slingshot and took aim. In obedience to his nod of command, I did the same, though with terrible twinges of conscience, vowing to myself that I would shoot as soon as he did. At that very moment the church bells began to ring, mingling their music with the songs of the birds and the sunshine . . . for me it was a voice from heaven . . . ever since then, when the Passiontide bells ring out to the leafless trees and the sunshine, I reflect with a rush of grateful emotion how on that day their music drove deep into my heart the commandment: "Thou shalt not kill" . . .*
>
> *From experiences like these that moved me deeply and often made me feel ashamed, there slowly grew up in me the firm conviction that we have no right whatsoever to inflict suffering and death on any of God's creatures unless an absolutely unavoidable necessity compels us to do so; that we should moreover realize the ghastliness of the fact that we do very often impose on them suffering and death from mere thoughtlessness* (Schweitzer 1949, 27–31).

For philosopher A. A. Luce (cf. Ch. 8 below), Schweitzer's description of his feelings constitutes a perfect description of sentimentality. And while Luce would allow that Schweitzer has every right to feel that way, he insists that he, Luce, has just as much right to continue fishing with a clean conscience, as long as he is not "cruel." In other words, according to Luce, Schweitzer, has made the mistake of elevating a subjective feeling to an ethical principle, with the result that Schweitzer's use of the word "cruel" is, from Luce's point of view, itself sentimental and lacking in precision. I show later how Schweitzer might defend himself against this charge.

In working out his approach to ethics, Schweitzer was conscious of being a revolutionary in several ways, and his inclusion of animals is one of these ways. He was quite aware, in this respect, of his predecessors in Indian and Chinese philosophy, and in the philosophy of Arthur Schopenhauer, but found either that they did not think the principles through to their end or that their thinking was rooted in a fundamentally negative attitude toward life. He was also aware that Jeremy Bentham included animals in the scope of his ethics, but writes, "Bentham, too, defends kindness to animals chiefly as a means of preventing the growth of heartless relations with other men, even though he here and there recognizes it as obviously right" (Schweitzer 1923, 297).

Any attempt to claim that contemporary nonsubsistence hunting in countries like the United States is rooted in an affirmation of and participation in the process of life, rather than being an expression of nihilism, as Joseph Wood Krutch claims, has to measure itself against Schweitzer's principle of reverence for life. In addition, the very words "reverence for life" give expression to a philosophy of life that should be attractive to thinkers looking for a radical approach to environmental issues today, especially those who advocate some form of "biocentric egalitarianism."[2] Biocentric egalitarianism—also called "biocentrism," "biospherical equalitarianism," and "ecological egalitarianism," among other variants—is the attempt to overcome the problem of "speciesism" by developing a principle of species equality (cf. French, 39). Reverence for life would seem to offer a very deep foundation for such attempts.

Schweitzer's influence has often been acknowledged in the animal protection movement, but his thought has rarely been taken up in recent environmental ethics. Indeed, J. Baird Callicott, one of the most important of contemporary environmental philosophers, writes, "Not to demean his biocentric ethic, Schweitzer was an amateur philosopher, less sensitive to the metaphysical constraints of Modernism than a professional might be" (Callicott, 251).[3] To be sure, Schweitzer was neither a naturalistically oriented analytic philosopher nor, in most senses of the term, a modernist; but then he did not want to be, and he had a deep understanding of just what it was that he did not want to be. Schweitzer studied theology and philosophy at the University of Strasbourg and later in Berlin. His dissertation was on Kant's philosophy of religion. Schweitzer's type of life-philosophy [*Lebensphilosophie*, see below] may be out of fashion, but he was no amateur; his opposition to modernism was itself philosophical. We cannot expect Schweitzer to have written in the 1920s the way he might have written fifty or sixty years later. We have to read him in his context and ask what he has to offer us in ours. For the moment, I simply hold onto the possibility that when he requires that a life view [*Lebensanschauung*] be optimistic and ethical, Schweitzer may provide a foundation for Krutch's accusation that sport hunting is nihilistic: "That world view is optimistic that places being higher than nothingness and thus affirms life as something valuable in itself.

From this relation to the world and to life results the impulse to raise existence, in so far as our influence can affect it, to its highest value" (Schweitzer 1923, 57).

ALBERT SCHWEITZER AND THE CRISIS OF CULTURE

Although most of my students today have never heard of him, Albert Schweitzer (1875–1965) was one of the great minds, as well as one of the great humanitarians, of the twentieth century. In 1912, Schweitzer left his position as professor of theology at the University of Strasbourg to become a mission doctor at Lambaréné in what was then French Equatorial Africa. By that time Schweitzer had written not only his highly original dissertation on Kant's philosophy of religion (published in 1899), but also a two-volume work on the Last Supper (1901), the book famous in English translation as *The Quest for the Historical Jesus* (1906), and a book on the psychiatric interpretation of Jesus (published in 1913). In addition, he was a well-known concert organist, and his book on Bach (1908) was groundbreaking. Later philosophical and scholarly works written in Africa include his *The Philosophy of Civilization* (1923), *The Mysticism of Paul the Apostle* (1930), *Indian Thought and its Development* (1935), and several studies on Goethe (1949).

While Schweitzer did not write in the context of environmental crisis, he shared the sense of living in a time of crisis and cultural collapse that was widespread in Europe in the late nineteenth and early twentieth centuries. *The Philosophy of Civilization*[4] opens with the words, "We are living today under the sign of the collapse of culture. The situation has not been produced by the war; the latter is only a manifestation of it" (Schweitzer 1923, 1). "Critique of culture [*Kulturkritik*]" was pursued by many intellectuals in Germany in the second half of the nineteenth Century and again in the aftermath of World War I, as thinkers sought to diagnose the cultural crisis of modernity that resulted in the Great War. Schweitzer thought that the immediate catastrophe was a symptom and effect of a deeper crisis that dates from the middle of the nineteenth century. With the failure of the Enlightenment and the failure of the attempts of the great philosophical systematists, such as Hegel, to root culture and ethics in a synoptic knowledge of reality as a whole, Schweitzer believed that European culture had lost its commitment to reason and to genuine progress. The diagnosis is carried out in a two-pronged analysis.

On the one hand, Schweitzer analyzes the general structures of the European life of his day (Schweitzer 1923, chapter 2). The main results of this analysis are that modern life is characterized by:

1. a diminished capacity for freedom, resulting from labor in the factory system;
2. a diminished capacity for thought, resulting from exhaustion and the need for distraction;
3. the fragmentation of knowledge into specialized fields and the resulting organization of human life such that "there is no call upon the whole man, only upon some of his faculties . . ." (*Ibid*, 13);
4. the anonymity of contemporary life, in which "we meet each other continually, and in the most varied relations, as strangers" (*Ibid*, 14);
5. the over-organization of public life, with the result that "modern man is lost in the mass in a unique way" (*Ibid*, 17).

This analysis is not particularly original. These themes were common in late-nineteenth and early-twentieth century sociology. There is also a strikingly similar analysis of "the present age" in Søren Kierkegaard's *Two Ages: The Age of Revolution and the Present Age*, which was completed in 1846.[5]

On the other hand, Schweitzer undertakes a comprehensive critical analysis of Western philosophy as an attempt to establish an optimistic and ethical stance toward life. Here he makes a strong distinction between a world view [*Weltanschauung*] and a life view [*Lebensanschauung*], the former based on theoretical, objective knowledge of the world, the latter on practical action, on willing (cf. *Ibid*, 76). Schweitzer defines a world view as "the sum of the thoughts that society and the individual concerning the essence and purpose of the world and concerning the place and destiny of humanity and of human beings in it" (*Ibid*, 49). A world view that is ethically optimistic includes the claim to *know* that the world itself is structured by and oriented toward ethical values. Western philosophy is the attempt to base an optimistic and ethical outlook on life [*Lebensanschauung*] on such an "optimistic-ethical view of the world"—that is, to base a practical life view on a theoretical world view—but Schweitzer's examination of this history (chapters 10–21 of Part II of *The Philosophy of Civilization*) is a critical survey of its failure. Schweitzer's response to this failure is radical:

> *My solution of the problem is that we must make up our minds to renounce completely the optimistic-ethical interpretation of the world. If we take the world as it is, it is impossible to attribute to it a meaning in which the purposes and goals of mankind and of individual human beings are meaningful. Neither affirmation of the world and of life nor ethics can be founded on what our knowledge of the world can tell us about the world . . . To understand the meaning of the whole—and that is what a world view is all about!—is for us an impossibility* (*Ibid*, 76).

While Schweitzer is not antiscience, he denies that science will ever give us access to a "meaning of the whole" on which we can erect an objective view of life and ethics.

SCHWEITZER AND LIFE-PHILOSOPHY

In view of this skepticism, how is an affirmative attitude toward life and the world, to say nothing of an ethics, possible? Schweitzer's answer to this question puts him squarely in the tradition of life-philosophy [*Lebensphilosophie*].[6] The *Historical Dictionary of Philosophy* distinguishes five different meanings of the word "*Lebensphilosophie,*" but it is mainly the fifth meaning that is relevant here:

> *The concept* "Lebensphilosophie" *gained a comprehensive use at the end of the nineteenth and beginning of the twentieth centuries among thinkers who began with the phenomena of inner life and its mental and historical-cultural expressions in order to overcome the rationalistic subject-object division. In doing so, they to some extent take up thinkers of the nineteenth century who developed a metaphysic starting from phenomena of the will. However, thinkers who tie the function of truth to its significance for the vital process have an influence on them to some extent.*[7] *Finally, there are dependences and references to pantheistic conceptions* (Pflug, 139).

Henri Bergson and Wilhelm Dilthey are the most famous representatives of this line of thought. Important predecessors are F. P. Maine de Biran, Arthur Schopenhauer, Friedrich Nietzsche, J. M. Guyau, and A. Fouillée. For Schweitzer, Schopenhauer, Nietzsche, and Guyau were the most important.

SCHOPENHAUER

Schweitzer's central concept in developing his ethics is "will to live," which he traces back to Schopenhauer. "Like Fichte, he [Schopenhauer] determines the essence of the thing in itself, which has to be assumed as underlying all phenomena, to be will, but not as in Fichte as will to have an effect [*Wille zum Wirken*] but, more immediately and more correctly, as will to live" (Schweitzer 1923, 236–237). But where Schopenhauer interprets this will to live as suffering and disappointment, and goes on to develop a pessimistic ethics of world renunciation (cf. *Ibid*, 237–240), Schweitzer demands that ethics be optimistic and life-affirming.

Schweitzer could take up one further aspect of Schopenhauer's thought in a positive manner, his inclusion of animals. Schweitzer summarizes Schopenhauer's ethics in the following succinct passage:

> *Ethics is pity. All life is suffering. The will to live which has attained to knowledge is therefore seized with deep pity for all creatures. It experiences not only the woe of mankind, but that of all creatures with it. What is called in ordinary ethics "love" is in its real essence pity* (*Ibid,* 239).

But while Schweitzer incorporates this concern for animal suffering into his reverence for life, he was generally careful not to give the word "pity" a central place, perhaps as a result of his reading of another representative of *Lebensphilosophie*, Friedrich Nietzsche.

NIETZSCHE

Schweitzer notes that Nietzsche begins to think under the influence of Schopenhauer, but then rebels against Schopenhauer's negative attitude toward life. On Schweitzer's reading of Nietzsche, starting with mature works such as *The Gay Science*, Nietzsche develops a philosophy of affirmation of life. The crucial element in this affirmation of life is

Nietzsche's emphasis on nobility: "What is noble? Nietzsche shouts to his age with hard words as the forgotten ethical question. Those who, as it echoed, were touched by the truth that was stirring and the anxiety that was quivering within it, have received from that solitary man all that he had to give to the world" (*Ibid*, 245). The crucial question, then, concerns what it is that constitutes "a higher affirmation of life" (*Ibid*, 245f.). While Nietzsche initially thought of higher affirmation of life in terms of the development of a higher spirituality or culture [*Geistigkeit*],[8] in his mature thought Schweitzer finds him taking a different direction. "Higher spirituality means, of course, the repression of natural impulses and natural claims on life, and is thereby in some way or other connected with negation of life. Higher affirmation of life, therefore, can consist only in the entire content of the will to live being raised to its highest conceivable power" (*Ibid*, 246). The element of negation of life in a higher spirituality is necessary, for Schweitzer, because focusing the will on the development and attainment of higher values requires that some natural impulses be controlled or negated.

This is what Nietzsche, according to Schweitzer, tried to avoid. But in his attempt to avoid subordinating the natural to the spiritual, Nietzsche falls to the opposite extreme:

> *Man [for Nietzsche] fulfills the meaning of his life by affirming with the clearest consciousness of himself everything that is within him—even his impulses to secure power and pleasure . . . Just in proportion as he emphasizes the natural, the spiritual recedes. Gradually, under the visible influence of the onset of mental disease, his ideal man becomes the "Overman,"*[9] *who asserts himself triumphantly against all fate, and seeks his own ends without any consideration for the rest of mankind* (Ibid, 246).

Schweitzer indicates his own direction in his summary judgment of Nietzsche:

> *In thinking out what affirmation of life means, Nietzsche is from the very outset condemned to arrive at the higher form of it by a more or less meaningless living out of life to the full. He wants to listen to the highest efforts of the will to live, without putting it in any relation to the universe. But the higher affirmation of life can arise only when affirmation of*

> *life tries to understand itself in affirmation of the world. Affirmation of life in itself, in whichever direction it turns, can become only intensified* [gesteigerte] *affirmation of life, never a higher form of it. Unable to follow any fixed course, it careens wildly in circles like a ship with its tiller firmly lashed.*
>
> *Nietzsche, however, instinctively shrinks from fitting affirmation of life into affirmation of the world, and bringing it by that method to development into a higher and ethical affirmation of life. Affirmation of life within affirmation of the world means devotion to the world. But this means that in one way or another negation of life appears within the affirmation of life. But it is just this interplay of the two that Nietzsche wants to get rid of, because it is there that ordinary ethics comes to grief . . . (Ibid,* 246–247).[10]

Schweitzer makes two points against Nietzsche here. First, he points out that an imperative to live out each and every one of one's impulses to the fullest without any ranking of those impulses, leads to an incoherent life. Second, he denies that such an approach, however much it is trumpeted as "affirmation of life," can lead to a genuine ethic, because an ethic must be not merely life-affirming but world-affirming. Such an affirmation of the world inevitably involves moments of negation of life, because devotion to the world requires that some impulses be resisted or overcome. The two points are closely interconnected in that the only way to escape the incoherence of the imperative of "living out of life to the full" lies in affirming the world, not merely the life one finds within oneself. A genuine ethic gives direction to life by being universal, world-affirming, in orientation. The ethical gives a determinate shape or direction to life, and this, for Schweitzer, is its spiritual dimension.

On the positive side, Nietzsche introduces the idea that "by living one's own life victoriously to the full, life itself is honored, and that by the enhancement of life the meaning of existence is realized" (*Ibid*, 247). His error, according to Schweitzer, lies in his one-sided emphasis on affirmation of life.

> *He is misled by the ethical element in affirmation of life into giving the status of ethics to affirmation of life as such. In doing so, he falls into the absurdities that follow from an exclusive affirmation of life, just as Schopenhauer falls into those of an exclusive denial of life . . .*

Both affirmation of life and negation of life are ethical for a certain distance; pursued to a conclusion, they become unethical . . . the ethical consists neither of negation of life nor of affirmation of life, but is a mysterious combination of the two (Ibid, 248).[11]

Fouillée and Guyau

In Fouillée and Guyau, Schweitzer finds two precursors of his own thinking in that they attempt to mediate between the negation of life of Schopenhauer and the affirmation of life of Nietzsche. Their weakness is that they attempt to ground their view of life in a philosophy of nature. This is doomed to failure even if it is in turn a form of life-philosophy. Both attempt to interpret ethics as simply the most fully developed form of evolution, which itself, according to Fouillée, is driven by "idea-forces [*idées-forces*]." Ethics is thus "a natural manifestation of the will to live" (*Ibid*, 255). In the ethical, the self perfects itself by devoting itself to the world. This is not a surrender of self, but rather its expansion.[12] Unlike Schopenhauer and Nietzsche, Fouillée and Guyau "hold on their course with a sure feel for the mysterious union of affirmation of the world, affirmation of life, and negation of life which constitutes ethical affirmation of life" (*Ibid*, 257). Their flaw is that they think that they can ground this ethic in a philosophy of nature, and here they fail in their attempt to show that nature is destined to become ethical. Ultimately, however, they recognize that this optimistic-ethical world view is a hypothesis that lacks any real certainty, and they are clear in their affirmation of their ethical position even if the hypothesis should have to be abandoned. "With these sentences there is announced from afar the disappearance of the optimistic-ethical interpretation of the world" (*Ibid*, 258). And with this the stage is set for Schweitzer's own ethics.

SCHWEITZER'S LIFE-PHILOSOPHY: REVERENCE FOR LIFE AS THE RATIONAL MYSTICISM OF REALITY

As I have already shown, Schweitzer abandons any attempt to ground ethics in an optimistic-ethical interpretation of the world: life view [*Lebensanschauung*] is no longer to be rooted in world view [*Weltanschauung*]; volition is no longer to be guided by pure knowledge. Yet Schweitzer thinks that European culture made a fatal error when it responded to the failures of the Enlightenment attempt to develop a rational optimistic-ethical world view by turning away from reason itself. The result is that an insidious form of pessimism has infiltrated a culture devoted to progress, a pessimism that is based on the failure of material-technical-scientific progress to provide for a genuine progress of humanity.

> *Pessimism is degraded will to live . . .*
>
> *It is where pessimism is at work in this anonymous fashion that it is most dangerous to culture . . . Thus the unavowed mixture of optimism and pessimism in us has the result that we continue to approve the external blessings given us by culture, things that to thinking pessimism are a matter of indifference, while we abandon that which alone it holds to be valuable, the pursuit of inner perfection* (*Ibid*, 97–98).

For Schweitzer, the solution to this dilemma is to abandon the attempt to develop an optimistic-ethical world view based on our knowledge of the world, and to turn directly to the development of an optimistic-ethical view of life. He starts not with objective knowledge about life (as in forms of life-philosophy oriented toward biology) but with the will to live as each individual experiences it immediately in his or her self.

"Deepened affirmation of world and life consists in this: that we have the will to maintain our own life and every kind of existence that we can in any way influence, and to bring them to their highest value" (*Ibid*, 278). But how does one get from the immediate experience of the will to live within oneself to this kind of deepened affirmation of life that encompasses every kind of existence? Schweitzer's commitment to Enlightenment rationality leads him to demand that the ethical be "a necessity of thought" (*Ibid*, 108; cf. 277). But as I show above, his analysis of the history of ethical thinking leads him to abandon the Enlightenment dream of an objective-ethical view of the world: the fundamental principle of ethics cannot be based on abstract or pure thought.

> *In the old rationalism, reason undertook to fathom the world. In the new it has to take as its task to gain clarity about the will to live that is in us. We thus return to an elementary*[13] *philosophizing that is once more concerned with the questions of*

world- and life-view that immediately move human beings, and establish and keep alive the valuable ideas that are in us. It is in a conception of life that is dependent on itself alone and seeks to come to terms with knowledge of the world in a straightforward manner, that we hope to find once more power to attain to ethical affirmation of world and life (*Ibid*, 277).

But this is also the point at which Schweitzer becomes explicitly mystical. When I[14] experience the will to live within me, I am in contact with something much deeper: "Reverence for life means to be in the grasp of the *infinite*, inexplicable, forward-urging will *in which all Being is grounded*" (*Ibid*, 283; my emphasis). This mystical experience (expressed in the italicized words) is crucial, but for Schweitzer it does not contradict the commitment to rationalism, because he thinks that mysticism is a "necessity for thought." "Every world- and life view that is to satisfy thought is mysticism. It must seek to give to the existence of man such a meaning as will prevent him from being satisfied with being a part of infinite Being in merely natural fashion, but will make him determined to belong to it inwardly and spiritually also, through an act of consciousness" (*Ibid*, 301). Whence this necessity for thought?

It is no accident that the chapter in which Schweitzer introduces his principle, the chapter entitled "The Foundations of Optimism Secured from the Will to Live," has two sections, one negative and the other positive. Failure to pay equal attention to the former as well as the latter may lead one to underestimate the complexity and power of Schweitzer's thought. The first section is a generalized statement of the argument Schweitzer had already developed in his critique of Nietzsche. It situates us, as it were, in the most elementary context of life and presents us with an either/or. Given, he wonders, the natural will to live each finds within him- or herself, what are the fundamental possibilities that arise "when thought awakens," thought that turns what had been natural and "self-evident" for prereflective life into a problem: "What meaning will you give your life? What do you mean to do in the world?" (*Ibid*, 278). The first possibility leads to shipwreck in the following manner, which passes through two stages.

I am will to live, but it takes only a little reflection to see that, as seductive as life is, it is full of disappointments.

> *What is spiritual* [das Geistige] *is in a dreadful state of dependence on our bodily nature. Our existence is at the mercy of meaningless events and can be brought to an end by them at any moment. The will to live gives me an impulse to action, but the action is just as if I wanted to plough the sea and sow in the furrows* (*Ibid*, 279).[15]

This motivates pessimism, but pessimism rooted in the will to live is inconsistent (*Ibid*, 280). I remain will to live, and this tension gives rise to a characteristic attempt to live one's life: "There arises an unthinking will to live that lives out its life trying to snatch possession of as much happiness as possible and that wishes to do something active without having made clear to itself what its intentions really are" (*Ibid*, 280).[16] Rather than devoting myself to action, as in the first stage, I live in the moment and grab at whatever happiness it may offer. Schweitzer's analysis of this form of life is a mixture of outside/inside perspectives, but it is the suggested immanent analysis that is most important. Such a will to live becomes "a kind of intoxication" (*Ibid*, 280) at the delight and beauty offered by life in the world, but then it once again comes up against "the suffering they discover everywhere within it" (*Ibid*, 281).

> *Now they see once more that they are drifting like shipwrecked men over a waste of waters, only that their boat is at one moment raised aloft on mountainous waves and the next sinks into the valleys between them, and that now sunbeams, and now heavy clouds, rest upon the heaving billows.*
>
> *Now they would like to persuade themselves that there is land in the direction in which they are drifting. Their will to live befools their thinking, so that it makes efforts to see the world as it would like to see it. It compels thought to hand them a chart which confirms their hopes of land. Once more they bend to the oars, till once again their arms drop with fatigue, and their gaze wanders, disappointed, from billow to billow . . .*
>
> *That is the voyage of the will to live which has abjured thought.*[17]
>
> *Is there, then, nothing that the will to live can do but drift along without thought, or sink in pessimistic knowledge? Yes,*

> *there is. It must indeed voyage across this boundless sea; but it can hoist sails, and steer a definite course* (*Ibid*, 281).

In Kierkegaard's terms, the will to live must choose itself. (Schweitzer's brief sketch of the despair (Kierkegaard's word) that develops when the will to live dedicates itself to beauty and pleasure in the moment reads like a book-jacket description of Volume I of Kierkegaard's *Either/Or*.) There are two important things to note here. The first is that when the will to live begins to reflect seriously, when "thought awakes," it is already situated and already in the service of the will to live itself. The second is that this situation is dialectical: the question is whether thought can give meaning to life, and we know in advance that failure amounts to existential shipwreck. Life has a stake in a positive outcome. There is nothing disinterested about such reflection, and it should neither be read nor evaluated in a "disinterested" manner.

In the second section of the chapter, Schweitzer unfolds the formal structure of this positive alternative to despair: a "deepened affirmation of world and life" (*Ibid*, 278). Rather than depending on knowledge of the world, the reef that shattered the first voyage of the will to live, this approach is based on "the will to live that becomes cognoscente of itself" (Schweitzer 1923, 281). The structural properties of any positive alternative to despair are as follows (cf. *Ibid*, 282–283):

1. to be true to the will to live (which pessimism abandons), the basic existential task, which immediately requires the will
2. to follow the natural impulse to ennoble my life and raise it to a higher power, and
3. to live life to the full in the sense of realizing the highest perfection of life.

These three are, Schweitzer says, given naturally in the will to live, *any* will to live.[18] But in the will to live that is given to itself as a problem, which has to come to clarity about itself and about the necessity of choosing how it is to live, additional tasks appear. Clarity about the necessity of choice leads it

4. to attain freedom from the world, in the sense of striving to replace natural impulse with considered choice, and once this has been achieved,

5. "the craving for perfection is given in such a way that we aim at raising to their highest material and spiritual value both ourselves and every existing thing that is open to our influence" (*Ibid*, 282).

and this just *is* the experience of

6. joining "in pursuing the aims of the mysterious universal will of which I am a manifestation" (*Ibid*, 283).

This pursuit is the expression of reverence for life. It is both an act of the will and an act of knowledge: an act of the will to live that has come to clarity about itself on the basis of knowledge of my will to live.

> *Thought is the interaction [*Auseinandersetzung*] between willing and knowing that goes on within me. Its course is a naïve one if the will demands that knowledge show it a world that corresponds to the impulses that it carries within itself, and if knowledge attempts to satisfy this requirement. This dialogue, which is doomed to produce no result, must give place to one of the right kind, in which the will demands from knowledge only what it knows* (*Ibid*, 308).

Just as the principle of reverence for life came to Schweitzer only when he was himself drifting aimlessly in his thinking about the problem of a culture with ethical depth and energy, the principle itself must be evaluated from the point of view of the will to live as it is lived, not from some "objective" or neutral point of view.

By the same token, "All true knowledge passes over into experience" (*Ibid*, 308) that yields "an inward attitude toward [*innerliches Verhalten zu*] the world and fills me with reverence for the mysterious will to live that is in all things" (*Ibid*, 309). In short, the fact that knowledge produces an inward attitude and behavior toward the world, which Schweitzer thinks is identical with reverence for the infinite will to live, is the criterion of the truth of this knowledge. The very meaning of truth is pragmatic in this existential sense: "truth" is a function of its significance for the will to live. This does not take away any of the rigor of the critical evaluation, but it is important to understanding the kind of rigor required.[19] The task is that of remaining true to the will to live I find within me. Beginning with "the most immediate and comprehensive fact of consciousness, which is: 'I am life that wills to live, in the midst of life that wills to live'" (*Ibid*, 309), thought discovers that affirming my own will to live requires an affirmation of that life that surrounds me. Such a deepened affirmation *is* reverence for life: "Reverence for life means being seized by [*Ergriffensein von*] the infinite, inexplicable, forward-urging Will in which all Being is grounded" (*Ibid*, 283).

Schweitzer's mysticism is the result of the existential struggle of the will. It is a "necessity for thought" for both negative and positive reasons. Negatively, Schweitzer claims that *only* such a mysticism can provide a way of overcoming the aimlessness of existence. An ethical "*mysticism* of reality" (*Ibid*, 305) is necessary for an optimistic view of life because the attempt to develop an objective, optimistic-ethical view of the world as a foundation for ethics fails. A "mysticism of reality" is necessary because traditional mysticism, dealing as it does with abstractions, tends to become "supra-ethical" (*Ibid*, 302–305). The mysticism of reality does not take the mystic to an Absolute beyond the phenomena of this world, but rather tries "to get its experience in living nature" (*Ibid*, 305). "There is no essence or substance [*Inbegriff*] of being, only infinite being in infinite manifestations" (*Ibid*, 305). Positively, in such a mysticism my will to live attains the truth about itself by becoming "true to itself" (*Ibid*, 282). "If knowledge answers solely with what it knows, it is always teaching the will one and the same fact, namely, that in and behind all phenomena there is will to live" (*Ibid*, 309). This is indeed a higher form of knowledge. "My knowledge of the world is a knowledge from outside, and remains forever incomplete. The knowledge derived from my will to live is direct, and takes me back to the mysterious impulses of life as it is in itself" (*Ibid*, 282).

Thus Schweitzer's mysticism consists in experiencing the will to live within me as an expression of the infinite will to live that underlies all phenomena of the world, and this experience is made possible (and necessary) by my devotion to raising all life to its highest value. Once I have experienced myself and the world in this manner, affirmation of my life becomes affirmation of life itself, and with this comes the most basic experience of responsibility. "The devotion of my being to infinite being means devotion of my being to all the manifestations of being that need my devotion, and to which I am able to devote myself" (*Ibid*, 305). This is the ethics of reverence for life. "Ethics arises in that I think out the full meaning of the affirmation of the world that is naturally given along with the affirmation of life in my will to live, and try to realize it" (*Ibid*, 307). In other words, if I become "genuinely thoughtful [*wahrhaft denkend*]," I become ethical (*Ibid*, 306) and mystical. If I thoughtfully will my own flourishing, I cannot but will the flourishing of life itself.

> *Ethics consists, therefore, in my experiencing the compulsion to show to all will to live the same reverence as I do to my own. There we have given us that basic principle of the moral that is a necessity of thought. It is good to maintain and to advance life; it is bad to destroy life or to obstruct it* (*Ibid*, 309).

Schweitzer recognizes that this principle goes far beyond the scope of traditional ethics because now "Ethics is responsibility without limit towards all that lives" (*Ibid*, 311). It is not based on compassion [*Mitleid*], because reverence involves concern not only for suffering, but also for "the urge for self-perfecting" found in all life (*Ibid*, 311).[20] The word "love" is more comprehensive, but Schweitzer thinks that it is limited by its erotic connotations. Ultimately, only the term "reverence [*Ehrfurcht*]" is adequate. The principle of reverence for life integrates two aspects of traditional ethics that Schweitzer calls "the ethics of devotion [*Hingebung*]" and "the ethics of self-perfecting [*Selbstvervollkommnung*]" (*Ibid*, 296).

Traditional ethics tends to emphasize either the one or the other, and to neglect or have trouble integrating the respective other. For example, social utilitarianism deals only with the devotion of person to person and is thus not universal, whereas the ethics of self-perfection is universal, but has trouble giving content to the concept of duty, even when it is world- and life-affirming (cf. *Ibid*, 288). "If the ethic of devotion, therefore, is to agree with the ethic of self-perfecting, it too must become universal, and let its devotion be directed not only towards the individual human being and society but also somehow or other towards all life whatever that comes to light in the world" (*Ibid*, 296). This is just what most traditional ethics has refused to do. Schweitzer notes, "European thinkers watch carefully that no animals run about in the fields of their ethics" (*Ibid*, 297). An ethic based on reverence for life, once one has recognized the will to live within me as an expression of a universal will to live, is not tempted by these traditional prevarications, because it recognizes in human beings "a mysticism of ethical union with Being" (*Ibid*, 309). This would be a traditional form of abstract mysticism if it were not rooted in the reality of our most immediate experience and if it did not immediately have concrete consequences.

REVERENCE FOR ANIMAL LIFE

In a passage that is important for the concerns raised in this book, Schweitzer contrasts the manner in which the universal will to live manifests itself in me (or us) with the way it appears in its other, nonhuman (or nonreflective) manifestations.

> *In my will to live, the universal will to live experiences itself otherwise than in its other manifestations. In them, it shows itself in a process of individualizing that, so far as I can see from the outside, is bent merely on living itself out to the full, and in no way on agreement or harmony [*Einswerden*] with any other will to live.*[21] *The world is a ghastly drama of the self-estrangement [*Selbstentzweiung*] of the will to live. One existence makes its way at the cost of another; one destroys the other. One will to live merely exerts its will against the other and has no knowledge of it. But in me, the will to live has come to know about other wills to live. There is in it a yearning to arrive at unity [*Einheit*] with itself, to become universal.*
>
> *Why does the will to live experience itself in this way in me alone? Is it because I have acquired the capacity of reflecting on the totality of being? Where does this evolution that has begun in me lead?*
>
> *To these questions there is no answer. It remains a painful enigma for me to live with reverence for life in a world that is dominated by creative will that is also destructive will, and destructive will that is also creative will.*
>
> *I can do nothing but hold to the fact that the will to live in me manifests itself as will to live that desires to come to harmony [*eins werden will*] with other will to live* (*Ibid*, 312).

This is Schweitzer's most basic statement of the meaning of reverence for life.[22] Reverence for life demands a harmony of all individual will to live, a harmony that would overcome the self-estrangement of the will to live. This is what it means to devote one's life to life.

Thus, in the ethics of reverence for life, the status of animals is recognized from the very beginning, in a very strong way. And while reverence encompasses compassion, devotion to life is based on neither pity nor compassion, but on a positive ideal of the perfection of life.

> *If I save an insect from a puddle, life has devoted itself to life, and the self-estrangement of life is overcome. Whenever my life devotes itself in any way to life, my finite will to live experiences agreement [*Einswerden*] with the infinite will in which all life is one. I enjoy a feeling of refreshment that saves me from languishing in the desert of life.*
>
> *I therefore recognize it as the destiny of my existence to be obedient to this higher revelation of the will to live in me. I choose for my activity the removal of this self-estrangement of the will to live, so far as the influence of my existence can reach. Knowing now the one thing needful, I leave the enigma of the universe and of my existence in it undecided* (*Ibid*, 313).

Once again, the fundamental existential nature of Schweitzer's thinking is obvious. The proof of the principle of reverence for life is the existential refreshment or comfort I find in acting on the principle, and in the fact that this (and only this) can save me from despair.

Schweitzer's actual discussion of relations between humans and animals is very brief in *The Philosophy of Civilization*. In addition to the passage just cited, he devotes about two pages to this issue (*Ibid*, 318–319). He has already emphasized that reverence for life commits one to protecting and enhancing all life that is within the compass of one's influence. But this cannot be the whole story, because any action that overcomes the self-estrangement of life is limited and temporary, as is the resulting harmony or "oneness." As finite beings, we find ourselves caught up in the world in which the will to live is divided against itself. As much as we devote ourselves to protecting and enhancing the life around us, we at the same time inevitably harm and destroy life around us. We cannot escape this necessity. Schweitzer asks how it is possible to lead a concrete life of reverence for life beyond the idealism of the principle of reverence for life. If the principle cannot help us make practical sense of this situation, it is useless. But this is not the case.

> *Whenever I injure life of any sort, I must be quite clear whether it is necessary. Beyond the unavoidable, I must never go, not even with what seems insignificant. The farmer who has mown down a thousand flowers in his meadow as fodder for his cows must be careful on his way home not to strike off the head of a single flower by the roadside for his insipid amusement, for he thereby commits a wrong against life without being under the pressure of necessity* (*Ibid*, 318).

Schweitzer is making two points here. First, one is permitted to do what one has to do. Injury to life is inconsistent with the ideal of reverence for life, and would not exist in a world in which the will to live is in harmony with itself. But we do not live in such a world, and while our actions can improve the world, they cannot completely heal the rift in the will to live once and for all, nor can they free us from being caught up in that division. Second, this permission extends only to the limits of the unavoidable.

Schweitzer's main example in this context is the use of animals in medical research. Following the principle of necessity, he demands that such research be in fact necessary for producing a result valuable for human beings (he does not mention veterinary medicine) and that the pain inflicted be mitigated as much as possible. There is no blanket condemnation of such experimentation, but this is not surprising, because Schweitzer makes it clear that he opposes the passivity of a one-sided ethics of self-perfecting or self-purity. Devotion to life requires more than just keeping one's hands clean.[23] One has to look at the good and at the harm, and decide for oneself whether one can justify the action. Ethics for Schweitzer must be activist ethics that is engaged in the world under the guidance of ideals rooted in reverence for life. Life is not only to be protected, it is also to be improved, and this requires that we be prepared to make hard decisions. The path of "leave them alone" is one of a false passivity. But, Schweitzer continues, the fact that experimentation on animals has yielded important medical advances establishes a special solidarity with and obligation toward animals beyond the general principle of reverence for life. "By helping an insect when it is in difficulties, I am only attempting to cancel part of man's ever new debt to the animal world" (*Ibid*, 318).

This leads to the most impassioned passage in this section.

> *While so much ill-treatment of animals goes on, while the moans of thirsty animals in railway cars sound unheard, while so much brutality prevails in our slaughter-houses, while animals have to suffer painful death from unskilled hands in our kitchens, while animals have to endure intolerable treatment from heartless men or are left to the cruel play of children, we all share the guilt* (*Ibid*, 319).

The ethics of reverence for life, by making us aware of our debt to animals, "makes us join in keeping on the lookout for opportunities to bring some sort of help to animals, to make up for the great misery that men inflict on them, and thus for a moment to step out of the incomprehensible horror of existence" (*Ibid*, 319). These are moments of agreement, of harmony, of oneness, not only with the life we aid, but also with the infinite will to live. They are moments of penance for the guilt human thoughtlessness brings upon each of us.

In one sense, this emphasis on the issue of "necessity" is by itself enough to clear the Schweitzer of *The Philosophy of Civilization* of the *ad hominem* charge of sentimentality. For all his deep compassion for the suffering of animals, Schweitzer does not allow himself to be seduced into a knee-jerk reaction that would produce a subjective good conscience while avoiding responsibility for making hard choices in the real world. (Invoking the distinction suggested earlier, I call the former "compassion" and the latter "pity.") This does not yet, however, demonstrate that there is not a form of objective sentimentality at work in the principle of reverence for life. I investigate this question in the next chapter.

SUBJECTIVITY AND NECESSITY: AGAINST RELATIVISM IN ETHICS

While Schweitzer is very quick and sure in his prohibition of the "insipid amusement" of beheading flowers (*Ibid*, 318), which is condemned for its thoughtlessness, he is very slow to condemn individual practices that might result from an individual's understanding and appropriation of reverence for life in the concrete situations forced upon us by the world, and this is a striking feature of his ethics. Reverence for life is not the principle governing a series of specific duties such as vegetarianism or the prohibition on nonsubsistence hunting. Indeed, Schweitzer thinks that one of the reasons traditional ethics is so careful to exclude animals is that it demands that ethics provide us with universally valid judgments, and sees that there is hope of providing this only if ethics restricts its concern to the realm of "the interests of human society" (*Ibid*, 298). On such a model, ethical deliberation consists in reviewing the list of rules supplied by ethics and applying it to the particular situation.[24] For Schweitzer, there are four things wrong with such an approach:

1. it restricts ethics to the realm of human affairs;
2. it makes ethics objective, and thus loses the existential source of the ethical;
3. it makes ethical value relative by seeking objectively justified compromises (*Ibid*, 317);
4. it tries to free us from the inescapable reality of ethical conflict.

Schweitzer cuts this Gordian knot by denying that a genuine ethics should provide an objective standard that can be applied by anyone in the same way in any given situation. "It is certain . . . that true ethics are always subjective, that they have an irrational enthusiasm [rooted in the will to live's thrust to live itself to the full, cf. *Ibid*, 282] as the very breath of its life" (*Ibid*, 299).

There is an apparent contradiction or at least tension in Schweitzer's thought here. On the one hand he demands "a system of ethics that is a necessity of thought" (*Ibid*, 108). On the other hand he insists on the subjectivity of ethics. But the contradiction is more apparent than real. As I show above, the necessity involved in ethics, as in the mysticism that underlies it, is existential, not objective. As Kierkegaard's Johannes Climacus puts it, it is the subjective passion kindled by the despair of the existing individual that leads to the "necessity" of postulating God (Kierkegaard 1846, 200, ftn.). The same holds in Schweitzer for reverence for life. The necessity is itself subjective, not objective. But there is also another subjective aspect of ethics.

> *In ethical conflicts, a person can make only subjective decisions. No one can decide for him just where, on each occasion, the extreme limit of possibility for his persistence in the*

> *preservation and furtherance of life lies. He alone has to judge this issue by letting himself be guided by a feeling of the highest possible responsibility towards other life.*
>
> *We must never let ourselves become blunted. We are in the truth when we experience these conflicts ever more profoundly. The good conscience is an invention of the devil* (*Ibid*, 317-318).

The position articulated here is a form of what is often called the "ethics of authenticity" (cf. Charles Taylor). When we are caught up, as we inevitably are, in the interplay of life and death, "in the conflicts that arise between the inner compulsion to devotion and the necessary self-assertion" (*Ibid*, 316), reverence of life demands that we be thoughtful. What is wrong in a straightforward and unqualified sense is unnecessary and thus thoughtless harm and destruction. Beyond that, Schweitzer maintains that each has to find his or her own way—thoughtfully. This is, as it were, the ethical condition once ethics drops its false ambition to provide ethically satisfying compromises.

The problem with an ethics that objectively weighs competing interests and offers rules according to which one (lower) interest is to be sacrificed to another (higher) interest is that it reconciles us with the harm we inflict on life. In place of the absolute claims of reverence for life, we are given the watered down demands of an ethics that relativizes one claim to another, thus making harm and death morally permissible and even good.

> *The ethics of reverence for life does not recognize a relative ethic. It allows only the maintenance and promotion of life to rank as good. All destruction of and injury to life, under whatever circumstances they take place, it condemns as evil* (*Ibid*, 317).

The price for an ethics born of subjective passion is that it must live from such passion at every moment. However thoughtfully and thus authentically one tries to live, the ethics of reverence for life "compels one to decide for oneself in each case how far one can remain ethical and how far one must submit oneself to the necessity for destruction of and injury to life, and therewith incur guilt" (*Ibid*, 317).

GUILT

Thus, Schweitzer's position is that living as we do in the concrete process of life, enmeshed in "the incomprehensible horror of existence" (*Ibid*, 319), we are always guilty no matter what we do or how thoughtfully we do it. Beyond the guilt we incur by our thoughtless harm and destruction of animal life, there is another form of guilt that is an intrinsic part of any will to live that reflectively, and thus ethically, engages the world. We are called upon to live ethically by our very awareness of the will to live, and all that can be demanded of us is that we act in the full enthusiasm of the ethical, that is, of reverence for life. Yet the result of Schweitzer's broad concept of guilt is that we are guilty at every moment of our lives, no matter what we do. Every time I breathe, I incur guilt, so all I can do is try not to breathe thoughtlessly.

Whence the guilt? Schweitzer's ethics works on two levels, one absolute and one concrete.

Absolute – My will to live identifies with the infinite will to live that underlies all phenomena. As such, reverence for life wills the flourishing of all life, of all manifestations of the infinite will to live. It requires that one act strictly and only in the service of life, so as to remove the self-alienation of the will to live such that it, and I, arrive at unity. This is absolute, and as such it recognizes the traditional ethical principle that "ought" implies "can" only in a limited sense. "The devotion of my being to infinite being means devotion of my being to all the manifestations of being that need my devotion, and to which I am able to devote myself" (*Ibid*, 305). The final clause is a clear statement of a limited version of the "ought implies can" principle. I am not called upon to devote myself to beings to which I am unable to devote myself, beings that are beyond my sphere of influence, and as long as I stay within the scope of this principle, no guilt can arise if I do not devote myself beyond the limits of what I can do. But nothing in this sanctions harming or killing other beings. Even unavoidable harm or destruction of life incurs guilt.

Concrete – My will to live finds itself thrown into a world that is more than a multiplicity of manifestations of the infinite will to live. It is a world in which the infinite will to live is divided against itself. To live in such a world is of necessity to live at the cost of other living beings, even as I devote myself to enhancing life, and my will to live can only turn to ethics to ask how I should live in such a world. Absolute reverence for life, when it finds itself in this world, can only say that if we must harm other life, as we must, we should do so only under the compulsion of necessity, and this requires that we act thoughtfully. Here "must" implies "may," just as "ought" implies "can." It remains true that, in absolute terms, I ought never to harm other life, but I cannot live in this world in this manner. Because I must harm other life, reverence for life permits me to do so under the strict condition of thoughtfulness. Thoughtless harm incurs guilt.

Why divide this into two levels? Does this do violence to Schweitzer's thought? I think not. The reason why recognizing the split is important is that the first level leaves its trace in the second level. If I simply live in terms of the second level, as long as I act thoughtfully and only under the compulsion of necessity, I act ethically. On this level, moral guilt is incurred by thoughtless action. The fact that Schweitzer here recognizes a wider version of the "ought implies can" principle is clear when he writes, "Whenever I injure life of any sort, I must be quite clear whether it is necessary. Beyond the unavoidable, I must never go, even with what seems insignificant" (*Ibid*, 318). And yet, Schweitzer insists that no matter how thoughtfully I act, I inevitably incur guilt. And this guilt cannot be the result of thoughtlessness. It is rather the result of the first level, the absolute demand of reverence for life. Viewed in isolation, the second level can produce only a "relative ethic," one that seeks and accepts compromises under the pressure of necessity—and has a good conscience about it.

THE SERMONS ON REVERENCE FOR LIFE OF 1919

It is striking that, although he is so famous for his attitude toward animals, Schweitzer devotes only two pages to the issue in *The Philosophy of Civilization*. However, in his first working out of the principle of reverence for life, a series of twelve sermons preached at Saint Nicolai Church in Strasbourg in 1919, during the closing weeks of World War I, Schweitzer devotes the entire third sermon to the issue of "our conduct toward living creatures" (Schweitzer 1919, 23). The sermons were translated into English under the title, *A Place for Revelation: Sermons on Reverence for Life*. The first three sermons are worth discussing, both because he offers more detail concerning our conduct towards living creatures and because at certain points Schweitzer has not yet seen just how radically egalitarian his new ethics of reverence for life will turn out to be when thought through consistently.

In the first sermon, Schweitzer develops the principle of reverence for life by presenting briefly many themes that he later elaborates in *The Philosophy of Civilization*: we need a morality based on reason; while we do experience within ourselves a duality of reason and "the heart," rather than being at odds, they complement one another, each leading to reverence for life; reverence for life involves "a compassionate sharing of experiences with all of life" (*Ibid*, 11). The sermon culminates in the statement, "You ought to *share life* and *preserve life*—that is the greatest commandment in its most elementary form. Another, and negative, way of expressing it is this: You shall *not kill*" (*Ibid*, 12).

In the second sermon Schweitzer introduces a new duality between reverence for life as the fundamental commandment of ethics and nature that "knows no reverence for life" (*Ibid*, 15). Schweitzer interprets the cruelty of nature as an "enigmatic rupture in the will to live—life against life . . . Nature teaches cruel egoism . . ." (*Ibid*, 16). This leads to the question of how we can "reconcile God-the-power-of-nature with God-the-ethical-will, the God of love" (*Ibid*, 17). No solution to this dilemma is offered here, and in *The Philosophy of Civilization* Schweitzer insists that there is no theoretical solution, no objective world view that shows that objective reality contains a reconciling teleology. The implicit message is that the reconciliation comes not on the level of theoretical world view, but on the level of the imperative to action. Reverence for life "is the great event in the development of life. Here truth and goodness appear in the world. Light shines above the darkness" (*Ibid*, 16).

One aspect of the second sermon that is worth noting is that Schweitzer speaks a hierarchical language here that he tries to avoid in later texts. Thus, in speaking of the senseless cruelty of nature, Schweitzer writes, "The most precious life is sacrificed to the most ignoble. A child breathes in the tuberculosis bacillus. He grows and thrives, but suffering and a premature death will be his lot because these lowly creatures multiply in his vital organs" (*Ibid*, 15–16). Two points are worth noting here. First, in later writings Schweitzer is careful to avoid words such as "ignoble" and "lowly." Thus, in *Out of My Life and Thought* (1931), he writes,

> *The ethic of Reverence for Life is judged particularly strange because it establishes no dividing line between higher and*

> *lower, between more valuable and less valuable life. It has its reasons for this omission.*
>
> *To undertake to establish universally valid distinctions of value between different kinds of life will end in judging them by the greater or lesser distance at which they stand from us human beings. Our own judgment is, however, a purely subjective criterion. Who among us knows what significance any other kind of life has in itself, as a part of the universe?* (Schweitzer 1931, 235).[25]

But, second, while Schweitzer is more careful in his later writings, there is a dualism in Schweitzer's thought that he finds ultimately unavoidable, and that cannot be written off as being "purely subjective." Life that has becomes aware of itself as such, that knows itself, is the light in the darkness of the natural world. It "blazes up, and that with the purest light, when it is forced to feed on what it derives from itself" (Schweitzer 1923, 283), that is, when it affirms its reverence for all life. This is the source of the dualism Schweitzer sees in the "dreadful state of dependence" of the spiritual on the body (*Ibid*, 279). Thus, a certain hierarchy is unavoidable, that of "the lordship of spirit over the powers of nature" (*Ibid*, 57), but for Schweitzer it is not a hierarchy of privilege, but rather one of responsibility.

In the third sermon, Schweitzer raises the question of human conduct toward living creatures. The first question concerns the extent of the life for which we must have reverence.

> How far down does the boundary of conscious, feeling life reach? *No one can say. Where does the animal stop and the plant begin? And the plants: Is it possible that they feel and are sensitive even if we cannot demonstrate it? Is not every life process, right down to the uniting of two elements, bound up with something like feeling and sensitivity?*
>
> Then every being must be holy to us. *We may not destroy anything of it carelessly* (Schweitzer 1919, 25–26).

This is a curious passage. The first paragraph begins with a statement of skepticism, raises a question that the skepticism indicates may well be unanswerable, and ends with a rhetorical question that gives a positive answer to the initial question, thus springing over the skeptical answer. The beginning of the next paragraph draws the ethical conclusion to this panpsychism. Its virtue is that it bypasses the embarrassment evident in many animal rights/animal liberation theories, which for practical reasons have to specify the demarcation line between organisms that are morally considerable and those that are not. It also has the virtue of bringing the whole of the natural world into the scope of ethics. The rhetorical question is the bud that blossoms into Schweitzer's rational mysticism in *The Philosophy of Civilization.*

The ethical problem arises from the fact that although we realize our nature as life aware of itself, as life sharing the life of others in the attitude of reverence for life, we are also a part of that world governed by the "self-alienation [*Selbstentzweiung*] of the will to live" (*Ibid*, 26). Schweitzer insists that this self-alienation is theoretically unresolvable. The resolution can only be ethical, that is, practical:

> *But precisely because we do stand so clearly under the terrible law of nature, which permits living beings to kill other living beings, we must watch with anxiety that* we do not destroy out of thoughtlessness, *where we do not stand under any pressure of necessity. We must perceive every act of destruction always as something terrible and ask ourselves, in every single case, whether we can bear the responsibility as to whether it is necessary or not* (*Ibid*, 26–27).

In the rest of the sermon, Schweitzer offers a series of specific examples, and the range is fascinating. Considering the practice of decorating living spaces with cut flowers, he writes,

> *The sight is beautiful, certainly. We bring nature into the room. But nature in what condition? Dying nature! The flowers in the vase die prematurely in order to please you. The picture over which you rejoice is the picture of death! . . . One day the time will come when all children will read in schoolbooks up to which century people in naïve cruelty pleased themselves with dying flowers* (*Ibid*, 27).

While this, as Schweitzer is quite aware, sounds to most ears like a quirky sentimentality, it is consistent with the attitude of intense consciousness of shared will to live, where what we identify with is the impulse to life, not death. The nihilism Krutch detects at work in sport hunting now wafts toward us from the flower arrangement on the table!

That bull fighting should be deeply incomprehensible to Schweitzer is obvious. Of more interest are his comments on hunting. According to Schweitzer, hunting can be necessary not only for procuring food, but also for avoiding overpopulation of animals in a given area. In this respect, the craft of hunting is compared—favorably!—with the craft of the slaughterer. What is condemned by the ethics of reverence for life is "hunting for pleasure [*Vergnügen*]"[26] and hunting "as a sort of training ground for manhood" (*Ibid*, 28). Again, this is hardly an attitude driven by sentimentality, but rather a clear and logical explication of the implications of his principle.

More curious is the question, "In the battle of life against life that plays itself out in nature, may I choose sides? May I involve myself in it?" (*Ibid*, 29). That I must be partial to myself and my own needs, at least to some extent, is already clear. The issue here concerns, to take Schweitzer's example, killing a spider, not to protect myself or other human beings, but so that it cannot "torture and kill" insects. Here Schweitzer falls back on the position (that we have already seen in *The Philosophy of Civilization*) that moral principles do not give rise to a series of objective, specific obligations and permissions. "Here no universally valid decision can be given. Rather, you must deal with every individual case according to your conviction and to your conscience. Perhaps one time you will do one thing and another time another" (*Ibid*, 30). Schweitzer gives two examples from his experience in Africa. There, he did kill the hawks that came to eat the young weaverbird chicks in their nests built in the palm trees around Schweitzer's house, as their parents uttered cries of grief. "This sorrow gave me the right to slay the robbers" (*Ibid*, 30). But he did not kill a sleeping alligator. "That was the case even though I could imagine just what sort of devastation he would wreak at night among the fish. But I did not catch him in the act and did not want to take the guilt on myself should he, wounded, dive into the water and suffer there" (*Ibid*, 30). Schweitzer does not raise the possibility that the hawks were feeding their own young, who starved as a result of his intervention. Similar concerns could be raised about the alligator. But for Schweitzer, this is a matter of individual integrity. "These decisions can go either way. If only you act responsibly and according to conscience, and not thoughtlessly, you are justified [*im Rechte*]" (*Ibid*, 30).

The same thing would presumable hold for the decision to be a vegetarian: if it is consistent with one's inner appropriation of reverence for life, by all means be vegetarian. But do not pretend that this removes one from participating in the self-alienation of the will to live. What Schweitzer's understanding of reverence for life makes clear is that one cannot demand that others be vegetarian. James Bentley writes that Schweitzer "became virtually a vegetarian—though not without difficulty" (Bentley, 135). This is not surprising, but it is clear that Schweitzer himself did not demand that others be vegetarian. What is imperative is that one take personal responsibility for the deaths one causes, and also take care not to inflict unnecessary suffering. "In Africa, where one must do all of one's own slaughtering, I force myself whenever possible to be present in order to prevent all unnecessary pain to the animal" (Schweitzer 1919, 31). This is not quite a personal demand that he eat only meat from animals he has killed himself. The crucial demand is that one not thoughtlessly leave the slaughtering to inexperienced hands. A professional butcher who has mastered the craft of slaughtering in as humane a manner as possible is another story.

We are part of the world, and are thus subject to its "terrible law of the self-alienation of the will to live." Reverence for life cannot demand that we not participate in this world, because that would require suicide on the part of the pinnacle of life, of the light that shines in the darkness (*Ibid*, 16). This would be to shirk rather than assume the responsibility demanded by reverence for life. Nor does it demand that we pretend that we can somehow heal the rift. What reverence for life demands is that we never destroy life thoughtlessly. Destruction of life may be necessary, but for one with reverence for life, "It is for us always equally terrible, equally uncomfortable" (*Ibid*, 31).

SCHWEITZER AND ANIMAL RIGHTS/LIBERATION

Schweitzer's emphasis on solidarity with life itself, with all life of whatever form, indicates that he is not simply *extending* the moral status that traditional ethics recognizes for human being to nonhuman life. "Extension" is a strategy or path of thought that presupposes that one has the correct moral principle, the only outstanding issue being its scope of application. While Schweitzer had *felt* an instinctive devotion to life since his childhood, in his systematic thinking he arrives at reverence for animal life on the basis of a general principle that was unknown to traditional ethical thought, not by extending an acknowledged reverence for one form of life outside of me to other forms of life. Indeed, Schweitzer's rejection of objectivity in ethics makes it clear that he would consider any extension of traditional forms of human ethics to the broader realm of life to be fundamentally mistaken, producing the specific form of "relativism" that he rejects.

Schweitzer's rational mysticism of reverence for life is also much less dogmatic than the positions of Regan and Singer, typical representatives of the ethics by extension. And there are deep reasons for this. Many philosophers and non-philosophers conceive of ethics as a system of principles that when adequately elaborated ideally tells us what to do in every adequately described situation. Both Singer and Regan come close to this. It is not clear that either embraces it completely, but each elaborates a general principle that either can be applied directly in a given situation or enables the enunciation of secondary principles (Regan's "Minimax," etc.) that increasingly determine the correct course of action in a given situation. Probably neither embraces it completely, because each could claim that many or most situations allow for a variety of permissible actions, that is, neither obligatory nor forbidden. But such actions are ethically neutral, not the proper object of ethical decision.

Schweitzer's approach is quite different. He sets out and grounds his general principle—but the grounding is existential, not objective. If this principle has any content at all, it must imply that some actions are simply wrong, because they are inconsistent with the general principle. Thoughtless killing and hunting for pleasure are examples. But there is, for Schweitzer, a much broader field of actions that are properly objects of ethical deliberation and decision, actions for which the principle of reverence for life is quite relevant, but that the principle does not straightforwardly show to be either obligatory or forbidden. But to say that they are permissible in the sense of being morally neutral is not correct either. Vegetarianism is one such issue for Schweitzer, and he makes it clear that some forms of hunting fall into this class. Neither vegetarianism nor subsistence hunting fall into the class of morally neutral decisions, for which it simply does not matter which one chooses from the moral point of view. Such morally neutral decisions do not tell us something fundamental about the kind of person one is and chooses to be. But deciding to be vegetarian *out of* reverence for life is a quintessentially ethical decision, and does tell us something about the kind of person one is and chooses to be.

For Schweitzer, these are questions for which the principle of reverence for life does not yield a univocally right and a univocally wrong answer. In this sense, while Schweitzer's principle of reverence for life does have deep and genuine significance for the way we live our lives, it is the antithesis of what, from Schweitzer's point of view, can only be called the dogmatism of Singer and Regan.

HUNTING AND REVERENCE FOR LIFE

The question I now raise concerns just how far Schweitzer's condemnation of "hunting as a pleasure" goes. As I have shown in Part I, many hunters would challenge this description of their hunting, even when it is not subsistence hunting, or even the source of an important amount of the food consumed. Is there a form of contemporary hunting that is an authentic expression of reverence for life? Can one hunt as a form of devotion [*Hingebung*] to life?

For a serious inquiry into the status of hunting for the ethics of reverence for life, one must go beyond the mere fact that Schweitzer was not a hunter. In a different situation, he might well have been, and where the lines are to be drawn in deciding to hunt or not to hunt is, for Schweitzer, an individual and subjective decision. What is not individual and subjective is the attitude in which such hunting would be conducted. In Africa, Schweitzer believed that he had to do his own slaughtering, since there were no professional slaughterers available. But his formulation is not accidental: he "forces" himself to be present at what had to be a very uncomfortable and terrible event. This is not, I submit, just an expression of personal sentiment, but also a straightforward requirement of reverence for life.

The question here is whether there is any middle ground between this kind of "forced" participation in killing, with its reluctance (and relief were it to be freed from the burden of necessity), and the "hunting for pleasure" Schweitzer condemns out of hand. Here the categories of Stephen Kellert help. On the basis of a survey of hunters, Kellert constructed three broad categories of hunters:

Utilitarian/meat hunters
Dominionistic/Sport hunters
The nature (or naturalistic) hunter (Kellert, 1978a; for a more detailed presentation cf. Kellert, 1978b).

From Schweitzer's perspective, the first category comes within the scope of the subjective decision, not that very many utilitarian hunters actually hunt from the perspective of reverence for life. The point is that reverence for life gives room for this kind of hunting. In Ted Kerasote's summary of Kellert's category of the dominionistic/sport hunter, "They often lived in cities and savored competition with and mastery over animals in the context of a sporting contest. However, their knowledge of nature and wildlife was very low" (Kerasote 1994, 212). This category falls under Schweitzer's condemnation of hunting for pleasure or as training for manhood, and reverence for life is, I think, inconsistent with dominionistic/sport hunting. (In some literature concerning hunting, those falling into this category would be disparagingly called "shooters," not "hunters.")

The interesting case is that of the nature hunter. As Ted Kerasote summarizes this category, "their goal was to be intensely involved with wild animals in their natural habitats. Motivated by genuine affection for wildlife, they were faced with the paradox of killing the creatures they loved" (*Ibid*, 212). Hunters of this kind may find hunting deeply meaningful and enjoyable, but typically they cannot be accused of hunting for amusement, much less of killing for amusement or pleasure. Kerasote, who clearly belongs in this category, demands of himself a moment in which he can "apologize" before shooting an elk. On the other hand, he in no sense "forces" himself to hunt. Indeed, at the end of a successful stalk early in the season, he can decide not to pull the trigger precisely in order to continue to hunt. Hunting itself is a deeply meaningful activity for Kerasote, and indeed a deeply pleasurable experience. Does Schweitzer have room for this kind of involvement, this kind of meaning, within reverence for life? It is clear that he himself would not have participated in this, but to some extent this is a matter of his own personal sensibility and integrity, and is not our concern here.

It is clear that from the perspective of *Schweitzer's* principle of reverence for life, the answer here has to be a decisive "no!" While Kerasote's engagement is an expression of the way life and death are intertwined in our world, it has an affirmative meaning that is foreign to Schweitzer. Kerasote writes,

> *The question then becomes, Can a cultural being ethically participate in these natural cycles, cycles that may entail taking the lives of individual animals, animals who are as bright,*

> *as bold, and tenderly aware of sunshine and storm as we are? Can one be both cultural and natural?* (*Ibid*, 178).

Schweitzer's answer to this question is "yes," but it is a deeply qualified affirmative. For Schweitzer, we *have* to participate. Indeed, we are condemned to participation. So yes, we can take their lives, as long as we do so thoughtfully. But in doing so we perform acts that are evil, and we incur a correlative guilt. Thoughtful killing is ethically permissible, but metaphysically evil. To give an affirmative meaning to the experience would be to affirm evil. (I investigate the concept of the "nature hunter" further, along with the concepts of "guilt" and "remorse," in the next chapter.)

Chapter 3

Critical Examination of Schweitzer's Ethics of Reverence for Life

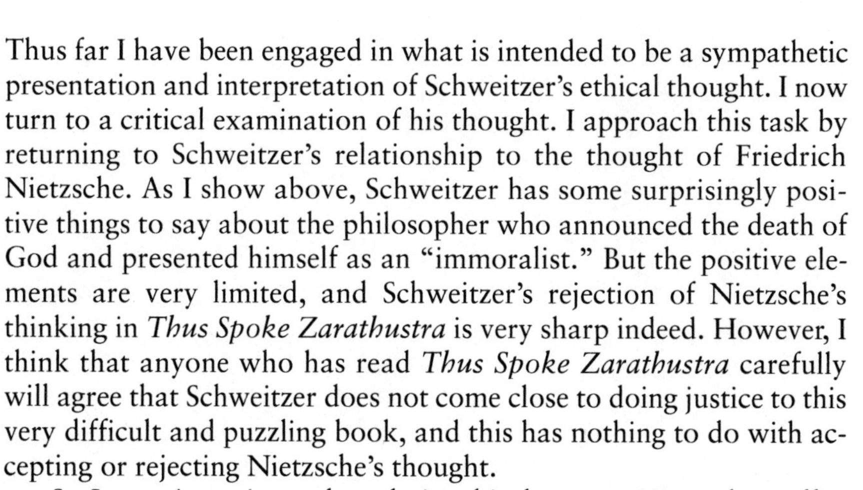

Thus far I have been engaged in what is intended to be a sympathetic presentation and interpretation of Schweitzer's ethical thought. I now turn to a critical examination of his thought. I approach this task by returning to Schweitzer's relationship to the thought of Friedrich Nietzsche. As I show above, Schweitzer has some surprisingly positive things to say about the philosopher who announced the death of God and presented himself as an "immoralist." But the positive elements are very limited, and Schweitzer's rejection of Nietzsche's thinking in *Thus Spoke Zarathustra* is very sharp indeed. However, I think that anyone who has read *Thus Spoke Zarathustra* carefully will agree that Schweitzer does not come close to doing justice to this very difficult and puzzling book, and this has nothing to do with accepting or rejecting Nietzsche's thought.

So I now investigate the relationship between Nietzsche's affirmation of life and Schweitzer's reverence for life, which is itself a form of affirmation. My goal is not a simple comparison, but a critical confrontation. In other words, I compare Schweitzer with precisely the Nietzsche he thinks is most clearly unacceptable, to see if Nietzsche raises questions that have to be taken seriously from the perspective of Schweitzer's principle of reverence for life, with philosophically important results.

SCHWEITZER AND NIETZSCHE: FORMS OF AFFIRMATION

Schweitzer couples his strong critique of much of Nietzsche with a strong approval of Nietzsche's critique of traditional Christian morality. He approves of Nietzsche's critique of the negation of life that is at the heart of much traditional morality, and he approves of Nietzsche's insistence that individual morality has primacy over social morality. To Schweitzer, the core of what Nietzsche has to offer the world is the scorching question "What is noble?" Yet when Nietzsche defines what a "higher affirmation of life" is, Schweitzer thinks that he stumbles. Rather than staying with the goal of a higher culture, Nietzsche tries to interpret affirmation of life without reference to spirituality or culture. Of course, Nietzsche in fact has quite a lot to say about culture. Schweitzer's claim is that Nietzsche makes the mistake of approaching culture from the perspective of a quantitatively *intensified* [*gesteigerte*] affirmation of life rather than pursuing the question of a qualitatively *higher* form of life (cf. Schweitzer 1923, 246–247). Such an intensified affirmation signifies living life to the full, but to Schweitzer this in turn can only mean living out every impulse one finds within oneself. On Schweitzer's reading of Nietzsche, "higher" can only mean range (all) and intensity (each to the fullest).

Early in Book I of *Thus Spoke Zarathustra*, Zarathustra speaks against "the despisers of the body," those who oppose the soul to the body and thus despise the body as that which holds down the soul (prominent examples would be Plato and Paul the apostle). But Zarathustra replies that the soul is a word for "something about the body" (Nietzsche, 34), and when the soul is opposed to the body, it is the body as a "sick thing" (*Ibid*, 33) that is speaking. The sick body is "angry with life and the earth" (*Ibid*, 35). The healthy body, in contrast "speaks of the meaning of the earth" (*Ibid*, 33). Against this kind of negation of life, Zarathustra teaches an affirmation of life: "A new will I teach men: to *will* this way which man has walked blindly, and to affirm it, and no longer to sneak away from it like the sick and decaying" (*Ibid*, 32). But if both the subordination of body to soul and the affirmation of the earth are actions of the body, on what basis can Nietzsche call on us to embrace one and reject the other? Nietzsche's answer is simple: the former is the body turned against itself, while the latter is the body striving to create beyond itself. Zarathustra says of himself "I am well disposed toward life" (*Ibid*, 41), not because he thinks that everything he finds is wonderful (on the contrary, there is much that arouses his disgust, and this is a continuing problem in *Zarathustra*). Rather, he interprets life as a restless process of self-overcoming, and it is this *process* toward which he is well disposed.

Zarathustra is well aware of the pain and suffering (and death) that are part of this process, and he recognizes that facing up to this pain and suffering poses a fundamental choice. Either the suffering that is an intrinsic part of life is taken to be an argument against life, a refutation of life (*Ibid*, 45), and a spur to negation of life and to the search for "after-worldly" solutions (cf. *Ibid*, 30–33; prominent examples are Christianity and Schopenhauer), or suffering is taken as a spur to creation. Such creation is the redemption of suffering in the sense that it gives meaning to suffering. "Creation—that is the great redemption of suffering, and life's growing light. But that the creator may be, suffering is needed and much change. Indeed, there must be much bitter dying in your life, you creators" (*Ibid*, 87). All creation is a process of self-overcoming, and all self-overcoming is painful. Even death, the ultimate and final negation of life, is taken up into this process and given meaning. Zarathustra calls on his followers to "Die at the right time!" (*Ibid*, 71). Death is part of the process of creation when it is "the death that consummates—a spur and a promise to the survivors" (*Ibid*, 71). Death should come when the creative project demands it, and in service to that project itself.

What is striking in this brief summary is the emphasis Zarathustra lays on *project* and *process*. Life is not valued for its mere aliveness, but rather for the self-transcending will that is found within it. "Life wants to climb and to overcome itself climbing" (*Ibid*, 101). Pain is not evil, nor is death evil. They are given meaning or value only when they are taken up either in the process of creation, which gives them a redemptive meaning, or in a reaction against life, which gives them a purely negative meaning as punishment and as that from which life wishes to be freed. But, Zarathustra warns, the latter is no longer death in the service of life and its creativity, but rather a pure will to nothingness—nihilism. Freedom from death is not the purity of life itself, but freedom from life.

The contrasts—as well as the similarities—between Nietzsche and Schweitzer are striking. At first glance, it would seem that the mature Nietzsche would have to refuse Schweitzer's mystical identification with an "infinite, inexplicable, forward-urging will in which all Being is grounded" (Schweitzer 1923, 283), but is this will really so different from Nietzsche's will to power?[1] Nietzsche could come close to affirming Schweitzer's words, but only in a very different interpretation. Schweitzer's will to live is set in opposition to suffering, destruction, and death. It is not that they are somehow injected from without, but they are not intrinsic to the will to live itself either. For Schweitzer, they are the result of the self-alienation of the will to live, whereas for Nietzsche they are part and parcel of the will to power—read here as Nietzsche's counterpart to Schweitzer's will to live. When Schweitzer writes, "The spiritual is in a dreadful state of dependence on our bodily nature" (*Ibid*, 279), he joins the ranks of the despisers of the body from Zarathustra's perspective. Nietzsche's version of "mystical identification" in willing eternal recurrence is intended to be the most radical form of affirmation possible. But rather than a mystical identification with an eternal will—Zarathustra insists that "no 'eternal will' wills" (Nietzsche, 166)—this is an identification with the process of becoming itself. This involves recognizing and willing the innocence of becoming,[2] which is absolutely foreign to Schweitzer.

Schweitzer's position is the result of his mystical identification with the infinite will to live. To identify with the will to live wherever it appears is to identify across all of the oppositions and conflicts between individual wills to live which we find in the empirical world. Because these individual appearances of the will to live are in constant conflict, the drive to universality and harmony is forced to posit a noumenal, infinite will to live that underlies each of the individual wills to life and their oppositions. The essential reality is the underlying will to live; the contingent reality that is the manifestation of that essential reality in the world of conflict around us. But as the manifestation of that universal, infinite will, we can act to enhance life and to alleviate suffering even as we find ourselves inextricably intertwined with the life-and-death nexus of the empirical world. Thus, we can move the worm on the concrete sidewalk back to the earth it needs, while seeing to our own butchering in order to make sure that it is done right.

This kind of reverence for life requires negation of life in the sense that devotion to infinite life, which requires devotion to life as it appears around me, inevitably requires that I must sacrifice some of my own life interests to my service to other life. This is precisely what Nietzsche denounces as weakness, and this is what Schweitzer criticizes him for. This is a real difference, but the chasm is not a broad as Schweitzer thinks it is. Schweitzer writes, "Gradually, under the visible influence of the mental disease that is threatening him, his ideal man becomes the 'superman,' who asserts himself triumphantly against all fate, and seeks his own ends without any consideration for the rest of mankind" (Schweitzer 1923, 246). But this completely misses Zarathustra's notion of the "bestowing virtue." Zarathustra contrasts an "all-too-poor and hungry" selfishness, a "sick selfishness" that greedily desires everything good for itself, a selfishness that is "degeneration," that says "Everything for me" (Nietzsche, 75), with a selfishness of strength, a selfishness that gives.

> *This is your thirst: to become sacrifices and gifts yourselves; and that is why you thirst to pile up all the riches in your souls. Insatiably your soul strives for treasures and gems, because your virtue is insatiable in wanting to give. You force all things to and into yourself that they may flow back out of your well as the gifts of your love. Verily, such a gift-giving love must approach all values as a robber; but whole and holy I call this selfishness* (*Ibid*, 74–75).

Zarathustra's point is that while he rejects a self-sacrificial giving based on pity and weakness, the creative and self-transcending will that chooses its own values and devotes itself to them is like a cup that, in filling itself, comes to the point that it overflows, bestowing its riches. This is a generosity based not on a self-limitation of life, but on its plentitude. But just as the overflow is the result of an engagement with life in all of its pain and suffering, so the riches it offers are not the cessation of pain and suffering (much less the end of death), but rather their redemption by offering new values and new goals toward which creative wills can strive. Zarathustra is generous, but never selfless.

But is this "bestowing virtue" just what Schweitzer rejects in Nietzsche, namely Nietzsche's demands for a complete affirmation of life, thus ignoring the element of denial that Schweitzer thinks is demanded by ethics and by any *higher* culture? After all, Zarathustra speaks of the bestowing virtue, but never of devotion (*Hingabe*). Schweitzer implicitly accuses Nietzsche of a lack of compassion. If this means pity (*Mitleid*) as Schopenhauer and Nietzsche understand it, then Nietzsche would proudly insist on it.[3] But Zarathustra is not without compassion. When he meets an unhappy youth who is desperately trying to mobilize his own self-overcoming and creativity, Zarathustra responds "It tears my heart" (*Ibid*, 43). He is full of compassion, but he does not attempt to take away the youth's pain and despair (this would be the response of pity): the youth must fight his own battles, although Zarathustra can warn him against some of the dangers he faces. If anything, Zarathustra challenges him on to more and more painful battles, but surely one can hear Schweitzer's term "devotion [*Hingabe*]" in his plea: "But by my love and hope I beseech you: do not throw away the hero in your soul! Hold holy your highest hope!" (*Ibid*, 44).

Schweitzer himself insists on an affirmative view of life. He summarizes this affirmation in *Out of My Life and Thought*:

> *If man affirms his will to live, he acts naturally and sincerely. He confirms an act that has already been accomplished unconsciously by bringing it to his conscious thought.*
>
> *The beginning of thought, a beginning that continually repeats itself, is that man does not simply accept his existence as something given, but experiences it as something unfathomably mysterious.*
>
> *Affirmation of life is the spiritual act by which man ceases to live thoughtlessly and begins to devote himself to his life with reverence in order to give it true value. To affirm life is to deepen, to make more inward, and to exalt the will to live.*
>
> *At the same time the man who has become a thinking being feels a compulsion to give to every will to live the same reverence for life that he gives to his own. He experiences that other life in his own. He accepts as good preserving life, promoting life, developing all life that is capable of development to its highest possible value. He considers destroying life, injuring life, repressing life that is capable of development to be evil. This is the absolute fundamental principle of ethics, and it is a fundamental postulate of thought* (Schweitzer 1931, 157).

As this passage makes clear, affirmation is devoted not to life as it is, but to life as it strives to be. Thus, devotion is directed toward preserving, promoting, and developing life. And while Schweitzer writes that "The philosophy of Reverence for Life takes the world as it is" (*Ibid*, 204), this does not mean that he affirms the world as it is, the world in which life and death, happiness and suffering, are so inseparably intertwined. Affirmation of life and affirmation of the world are not the same thing!

Ultimately, the deepest difference between Schweitzer and Nietzsche is that when Schweitzer calls for an affirmation of life, he means life as opposed to death, pure life freed from any intertwining with death. When Nietzsche calls for an affirmative relation to life, to its enhancement, it is to the *process* of life, to life as it occurs in the world. As a personified "life" tells Zarathustra, "Behold . . . I am *that which must always overcome itself*" (Nietzsche, 115). For Nietzsche, talk of affirming life can only mean the affirmation of *this* life. Schweitzer could never speak of the innocence of becoming, only (perhaps) of the innocence of the infinite will to live that grounds all being. No particular will to live, in this world in which the will to live is turned against itself, can be called innocent. As such, it is caught up in evil and, in the form of life that has become aware of itself as such, guilt. From Nietzsche's perspective, Schweitzer could be read only as falling back into the fundamental weakness and negation of life of the Christian tradition.

THE PROBLEM OF SCHWEITZER'S "MYSTICISM OF REALITY"

This is more than a simple difference of opinion (in which one of the persons involved is sliding into mental illness). Nietzsche's insistence that "life" can refer only to the concrete process of life, and not to some "afterworldly" reality that underlies the phenomenal world, raises a serious problem for Schweitzer. Schweitzer insists that his mysticism does not continue the tradition of abstract mysticism. For such mysticism, "the essence [*Inbegriff*] of being, the absolute, the world spirit, and all similar expressions, refer to nothing actual, but to something conceived in abstractions, which for that reason is also absolutely unimaginable" (Schweitzer 1923, 304). A mysticism committed to such an abstraction is "supraethical," because it loses contact with concrete reality. For Schweitzer, the only reality is the being that manifests itself in phenomena [*das in Erscheinungen erscheinende Sein*]" (Schweitzer 1923, 304). This being is the "infinite . . .will in which all being is grounded" (*Ibid*, 283, where "being" has to refer to phenomenal being). But what is this infinite will to live as distinguished from the concrete flow of phenomenal life in this world, in abstraction from the intertwining with death that makes life possible, in abstraction from concrete existence, in which, as Hegel notes, being and nothingness are intertwined?

Schweitzer objects that traditional abstract mysticism tempts one into an illusory "relation with the totality of being, that is to say, with its spiritual essence" (*Ibid*, 304).

> *Reality knows nothing about the individual being able to enter into connection with the totality of being. Just as it knows of no being except that which manifests itself in the existence of individual beings, so also it knows of no relations except those of one individual being to another. If mysticism, then, intends to be honest, there is nothing for it to do but to cast from it the usual abstractions, and to admit that it can do nothing rational with this imaginary essence of being . . . It must in all seriousness go through the process of conversion to the mysticism of reality. Abandoning all stage decorations and declamations, let it try to get its experience in living nature* (*Ibid*, 304–305).

Does Schweitzer's mysticism live up to these self-imposed standards? It may seem that it does, beginning as it does with my own will to live and coming to experience all reality as being alive, but the moment Schweitzer passes from speaking of "infinite being in infinite manifestations" (*Ibid*, 305), where infinite being is nothing other than the being of the manifestations, to speaking of being in the grasp of "the infinite, inexplicable, forward-urging will *in which all being is grounded* [*in dem alles Sein gegründet ist*]" (*Ibid*, 283; emphasis added) and of the will to live "*behind* and in [*hinter und in*] all phenomena" (*Ibid*, 309, emphasis added), he reaches beyond living nature to an abstraction. The goal of ethical life is to enhance life and raise it to its highest value. But when "life" is interpreted from the perspective of the infinite will to live *behind* all phenomena, the goal is pure life, freed from its intertwining with pain, loss, and death. And such a life, thought concretely, is death, death at least to organic life. It is a purely spiritual life, severed from the bonds of the body. The goal of an affirmative relation *to the world* has been missed.[4]

This can be put more concretely. The "highest possibility" of a wild animal is not some kind of pure life, but is precisely a function of its wildness, of the fact that its species has come to be what it is as a result of evolutionary pressures—climate, geography, predation, etc. To respect or have reverence for *this* life is to have reverence for its wildness, and this is the antithesis of an attempt to free it from that wildness. From this perspective, which is decidedly *this worldly*, actions such as Schweitzer's killing the hawks and his being tempted to kill the crocodile to keep it from wrecking havoc with the fish, is, as he acknowledges in a text written in 1952, "totally arbitrary"

(Schweitzer 1952b, 22–23), since taking sides for the fish is taking sides against the crocodile. As Ortega y Gasset writes,

> *One should not look for perfection in the arbitrary, because in that dimension there is no standard of measure; nothing has proportion nor limit, everything becomes infinite, monstrous, and the greatest exaggeration is at once exceeded by another* (Ortega y Gasset, 104).

Schweitzer, probably unintentionally, lays the ground for Cleveland Amory. Against his wishes, reverence for life has become arbitrary and thus sentimental, this time in an objectionable sense.

Another result of Schweitzer's commitment to an abstract "will to live" is that, while Schweitzer gives a great range for individual ethical decision, what he rules out is that bowing to the necessity of killing should have any affirmative meaning beyond that of providing necessary nourishment. He forced himself to supervise the slaughter of animals, bowing to necessity. But there is no room in Schweitzer's mystical reverence for (abstract) life for having this be a "sacramental" act in Thomas McGuane's sense (cf. Part I), a way of affirmatively participating in the web of organic life. To do so would be to cover over the evil and the guilt. If the intertwining of life and death is just a contingent, regrettable fact, this cannot be otherwise. In reverence for life understood in this manner, our lives become a penance for life, just as Schweitzer supervised the slaughter as a kind of penance for the fact that will to live is divided against itself.

When Schweitzer writes "That world view is optimistic that places being higher than nothingness and thus affirms life as something valuable in itself. From this relation to the world and to life results the impulse to raise existence, in so far as our influence can affect it, to its highest value" (Schweitzer 1923, 57), this is powerful. But when "life" is interpreted as infinite will *behind* phenomena, as pure being, it loses contact with the world. Whereas traditional mysticism is, Scheweitzer argues, supraethical, his own mysticism is hyperethical: we are guilty simply by virtue of the fact that we exist. If we could only perfect ourselves, we would free ourselves from being caught up with organic life, in which death and life are intertwined. We would be pure being, which, as Hegel notes, is conceptually indistinguishable from pure nothing.

QUESTIONS ABOUT HUNTING

If this critique of Schweitzer's hyperethical mysticism is correct, what would a reformed conception of reverence for life look like? Paul Taylor's concept of "respect for nature" appears to be an attempt to work out such a conception in the context of a more contemporary sense of environmental crisis. But before turning to a discussion of Taylor, I shall investigate the ways in which an approach like Schweitzer's, but freed of his abstract mysticism, allows us to pose questions concerning hunting. Schweitzer, unlike most traditional ethical theorists, allows a large range for personal authenticity, so the question "Is nonsubsistence hunting ethically permissible?" which is often taken to be *the* question, can be seen to contain several distinct questions. Failure to distinguish these questions makes a critical discussion of the issues very difficult.

Consider the following passage from Buddy Levy's *Echoes on Rimrock: In Pursuit of Chukar Partridge*. It begins with a question over dinner:

> *"I just can't understand how you consciously choose to spend your spare time chasing down and then shooting beautiful, innocent birds."*
>
> *She had a point. The birds I pursue are beautiful, which is one of the main reasons I love to watch them. I am not in a position to speak to their innocence . . . that's better left to avian theologians or the god of feathered creatures. "Well," I said, pausing. I wanted this to come out right. She had, after all, asked a very good question, one for which every hunter, I believe, should have an ever-evolving answer.*
>
> *I asked for more chardonnay. This might take awhile. Jim Harrison says in his foreword to Guy de la Valdéne's book* For a Handful of Feathers, *"hunting can be a good experience for your soul, to the degree that you refuse to exclude none of the realities of the natural world, including a meditation on why you hunt, perhaps an ultimately unanswerable question."*[5] *A paraphrased version of this quote was swirling around in my head as the white wine swirled around the lip of the glass, but I knew she wouldn't let me off that easily even though Harrison is right.*

> *"Are you a vegetarian?" I asked, and before she had the chance to respond I blurted out, "Those look like leather shoes you are wearing!" But I knew that the point of these questions had little to do with the point of hers. She wanted to know how I could kill something. She wanted to know why I killed chukars. I decided she deserved some fair answers, not a counterattack, which would lead us nowhere but our respective corners* (Levy, 128–129).

Levy then speaks to the way most of our meat comes from the supermarket via an anonymous process that allows us to forget its origin. He describes hunting for and killing chukar partridge, cooking and eating them, and then returning some of the bones "to the place that offered them life" (*Ibid*, 130). He describes his five-year-old daughter going hunting with him, helping him pluck the birds, watching him clean the birds.

> *She will respect life more deeply than most children do today because she has had the opportunity to witness death . . . Will this firsthand knowledge turn her into a heartless, thoughtless killer of harmless animals? I suspect not. I hope the conversations, our hours together plucking birds, of being out in wild country where the birds live, will develop in her a deeper appreciation for where some of her food comes from, how the chain of life works, and what her part is in the circle* (*Ibid*, 132–133).

His friend responds that "she simply could not condone it" (*Ibid*, 133).

This conversation is ostensibly about the ethical status of hunting, but it actually shifts from one question to another, without either of the participants being clear about this. Once it is freed from the hyperethical commitments of his abstract mysticism, Schweitzer's approach to ethics allows us to distinguish at least four different questions in this conversation:

1. Is nonsubsistence hunting ethically permissible in principle?
2. Does *my* appropriation of the principle of reverence for life allow hunting?
3. How can one kill such beautiful creatures?
4. Why do we hunt?

Clearly, these questions are interrelated. The person who asks the third question with genuine passion and incomprehension will almost certainly give a negative answer to the second question when subsistence is not at issue. An answer to the last question can help answer the third question, and will generally lead to an affirmative answer to the second question. And the answer to the first question may depend on the answer one can give to the fourth. But keeping these questions distinct is important.

I think that it has to be granted at the outset that if one's personal answer to the fourth question, "Why do I hunt?" is "I find it amusing," then it has to be considered immoral even for a reformed conception of reverence for life. It does not matter whether this means that it is amusing to kill animals or that the entire process of hunting, of which killing is a component, though an essential one, is amusing. Hunting for amusement is, I submit, inconsistent with any form of reverence for life. But to assume that all nonsubsistence hunting is a matter of amusement is to prejudge the issue. As I argued in Part I, James Swan's claim that "like his predecessors, the modern hunter hunts for meaning" (Swan, 144) suggests that "amusement" or "pleasure" does not begin to do justice to the experience.

In the conversation, Buddy Levy focuses on the fourth question "Why do we hunt?" His attempt to answer it leads to an answer to the third question, "How can one kill such beautiful creatures?" This in turn leads to an answer to the second question, "Does my appropriation of the principle of reverence for life allow hunting?" (Levy's word is "respect.") All of which presupposes that the first question, "Is nonsubsistence hunting ethically permissible in principle?," has *already* been answered in the affirmative. His friend starts with the third question, "How can you kill such beautiful creatures?" and her stumbling point is the fact that Levy finds the hunting, and the killing it involves, enjoyable: "How can going out on a weekend and killing things be enjoyable. I just don't get it" (*Ibid*, 131). Her question is not aimed at the professional butcher, and she herself is not vegetarian. Her focus is on enjoyment (as Schweitzer was concerned with amusement or pleasure); Levy's is on meaning. That the entire process of hunting, killing, cleaning, cooking, eating, and returning the bones is deeply enjoyable is of course part of the meaning of the actions. (I suggest in Part I that Richard Nelson's description of the entire process as being deeply "satisfying" is most adequate.) But killing itself, a moment within the entire process, is not by itself the final cause of the process. It is clear that Levy does not find hunting so meaningful and enjoyable because he enjoys killing.

In an earlier chapter of the book Levy describes the first time he went hunting for chukar with his father. When he kills his first chukar and holds the dead bird in his hand, he "marveled at its beauty" (*Ibid*, 5).

> *I looked at the bird's eye, at its beauty and wildness, and wished for a moment that the bird could still see, that I had not cut its life short with one blast of my shotgun. The living, breathing bird, just moments before in flight, now lay slumped in my hands, and I was not certain why. I held the bird out toward my father, wondering what to do next.*
>
> *"Nice shot, son." Then he added, "Beautiful, isn't it?"*
>
> *He slipped the bird into my game poke. I would feel its weight as we walked, the body bobbing against my tailbone. I had drawn the blood of a wild animal for the first time, and in my adolescence I felt, also perhaps for the first time, ambivalent. I was exhilarated and sad at the same time. It surprised me how quickly and instinctively, without hesitation, I had pulled up and fired. I had been deeply excited by the pursuit, and yes, even by the kill, but there was a twinge of guilt in my heart as I walked along. I think this is the moment at which one either becomes a hunter or does not—one's comprehension of the first kill, and how one reconciles it. I wiped the blood from the back of my hand and hurried off after my father. We paused at a fence line.*
>
> *"We'll eat chukar for supper tonight," he said* (*Ibid*, 5–6).

The contrast between this scene and Schweitzer's experience of hunting sparrows with his childhood friend is striking. Does "training ground for manhood" (Schweitzer 1919, 28) really capture the kind of lesson for life Levy finds here?

I suggest that Buddy Levy's account of why it is that he hunts is thoroughly consistent with what I call the "concrete" level of Schweitzer's reverence for life. This is the level of life as it is in the world we live in, in which life and death are inextricably intertwined. The question is not whether or not Schweitzer himself would have had a deeply empathetic understanding of Levy's attempt to articulate thoughtfully why he hunts, that is, why he is so passionate about it and why it is such a meaningful part of his life. There is room in Schweitzer's understanding of ethics for a great deal of "I just don't get it." This is the correlate of the room it gives for personal, thoughtful decision.

Levy himself is aware of this. He does not put forward the hunting life as a superior way of living. Speaking of Logan, his five year old daughter, he writes,

> *Logan has not yet fired a gun and possibly never will. The choice will be hers to make . . .*
>
> *Years from now, she'll have the choice: to fire on a particular bird, sacrificing its noble life for her evening meal, or to be content to let that bird keep flying, to simply marvel at its beauty, to commit that gorgeous, distilled moment to memory, to her own idea of what it means to be free. And I hope that if I have done my job, she will possess all the necessary information to make the correct decision, the choice that's right for her* (Levy, 134).

Sometimes, of course, "not getting it" is indeed a serious objection . . . against the one who does not or even cannot get it. Schweitzer's commitment to rationalism in ethics would, however, mean that no debate concerning ethical principles could be decided by "I don't get it." Levy's friend moves from "I don't get it," which belongs on the level of subjective decision, to "I can't condone it," which belongs on the level of ethical principle. For Schweitzer that would be a mistake. While the friend's "I just don't get it" is not necessarily an expression of sentimentalism, it becomes sentimentalism when moves to "I can't condone it."[6]

A wonderful example of someone viscerally opposed to hunting who is yet able to go some lengths to understand it is Terry Tempest Williams. Williams may be as pure an embodiment of Schweitzerian reverence for life as I have encountered.[7] A few years ago she quit her job as a naturalist in residence at the Salt Lake City Museum of Natural History because, as she said, she was tired of being surrounded by dead things. In her essay "Deerskin," she describes the excitement of her father and brothers when they discover deer tracks in the snow outside their house.

> *"Deer tracks . . ." my father said, touching them gently.*
> *"Deer tracks," I said. "So?"*
> *"Deer tracks," my brother restated emphatically.*

> *"Deer tracks," I said again under my breath. No, something was missing when I said it. Feeling out of place and out of touch, I went back inside and shut the door. Through the glass I watched the passion that flowed between my father and brothers as they spoke of deer. Their words went beyond the occasion* (Williams, 51).

Williams approaches an understanding of that passion by relating it to some aspects of the Navajo Deerhunting Way, "a blessing rite, a formula for corresponding with deer in the appropriate manner" (*Ibid*, 52). The Deerhunting Way includes stories of origin, in which the Deer Gods themselves teach the first divine hunters how to hunt. The result is, Williams comments, "a model for ecological thought expressed through mythological language" (*Ibid*, 55).

> *It is this kind of oral tradition that gives the Navajo a balanced structure to live in. It provides continuity between the past and the future. They know how to behave. Stories channel energy into a form that can heal as well as instruct. This kind of cosmology enables a person to do what is appropriate and respect the rights of others* (*Ibid*, 55).

Williams then connects the power of this kind of oral tradition with what she participated in and learned when she accompanied her father on a deer hunt when she was sixteen. She speaks of the "rituals" of the hunt: specific clothing, rubbing mink oil into boots and insect repellent into their skin, rising early to hunt and stalking the ridges late. Finally, around the campfire, she listened to the stories her father told.

> *It was only then that I realized a small fraction of what my father knew, of what my brothers knew about deer. My brothers had been nurtured on such tales, and for the first time I saw the context they had been told in. My education was limited because I had missed years, layers, of stories.*
>
> *Walk lightly, walk slowly,*
> *Look straight ahead*
> *With the corners of your eyes open.*
> *Stay alert, be swift.*
> *Hunt wisely*

In the manner of the deer.

I walked with reverence behind my father, trying to see what he saw. All at once he stopped, put his index finger to his mouth, and motioned me to come ahead. Kneeling down among the scrub oak, he carefully brushed aside some fallen leaves. "Deer tracks," he said.

"Deer tracks," I whispered (*Ibid*, 56–57).

The essay ends with these words, in which nothing is missing.

But this is not to say that Williams became a hunter. Far from it. As I understand what I have read of Williams, she feels a deep repulsion to hunting in our culture, which is not dependent on hunting animals for human subsistence. She is very passionate in this opposition, yet in this essay she clearly leaves a lot of room for an authentic, thoughtful, and deeply meaningful relationship to deer on the part of deer hunters. Her essay is not so much about hunting as about her own attempt to connect with the intensity and integrity she heard in her father's voice, even as the practice itself remained foreign, if not repugnant. But that understanding, mediated through insight into the web of experience and stories that constituted the hunting culture of her father and brothers, allowed her to walk "with reverence" behind her father, trying to see not only what he saw, but also how he saw it, more, to see in the way he saw it.

Part III

Paul Taylor's Ethics of Respect for Nature

Chapter 4

The Biocentric Ethics of Paul Taylor

Albert Schweitzer did not develop his principle of "reverence for life" as a response to a perceived environmental crisis. But his attitude of reverence for life could offer guidance to one seeking to reorient modern values in the face of the spreading destruction of the natural world. However, while his thought has been widely discussed in the animal protection movement and has been influential in those circles, it has tended to remain on the periphery, at best, of most environmental philosophy. But this does not necessarily mean that all of his ideas have been in eclipse.

In an essay entitled "The Ethics of Respect for Nature," published in 1981 and included in many if not most collections of works on environmental ethics since then, Paul Taylor argues that wild living things possess inherent worth and should be respected as such. "To say that it [a living thing or group of living things] possesses inherent worth is to say that its good is deserving of the concern and consideration of all moral agents, and that the realization of its good . . . [is] to be pursued as an end in itself and for the sake of the entity whose good it is" (Taylor 1981, 201). To recognize the inherent worth of living things and to give them their due consideration in this light is "respect for nature." Taylor gives a detailed presentation of his theory of environmental ethics in his 1986 book entitled *Respect for Nature.*

Taylor's approach has been enormously influential in the field of environmental ethics. Bryan G. Norton reviewed *Respect for Nature* in the journal *Environmental Ethics*, calling it "the standard against which future theories of environmental value will be judged" (Norton, 261). In particular, Taylor's principle of respect for nature has resonated through the approaches to environmental ethics that are generally grouped under the title "biocentric egalitarianism." It has been criticized by some, modified and developed by others. James Sterba has developed a revised version of Taylor's principle, one which Sterba thinks is less vulnerable to the charge of harboring a hidden anthropocentric bias (Sterba, 1998).

The kinship of Taylor's principle of respect for nature to Schweitzer's principle of reverence for life is clear. When Schweitzer writes, "Deepened affirmation of the world and of life consists in this: that we have the will to maintain our own life and every kind of existence that we can in any way influence, and to bring them to their highest value" (Schweitzer 1923, 278), this amounts to making recognition of inherent worth our principle of action. Schweitzer's notion of "*eins werden*," of harmony or becoming one with life outside of me, involves taking the good of other lives as something to pursue as an end in itself, and it finds its parallel in Taylor's claim that in a biocentric outlook on nature we gain "a sense of oneness with all other living things" (Taylor 1986, 115).

But Taylor's ethics is not just a restatement of Schweitzer. (Indeed, Schweitzer was not a direct influence on Taylor's thinking at all.[1]) In the first place, Taylor's thinking is much more ecologically oriented than is Schweitzer's. When Taylor speaks of human beings as "members" of the "Earth's Community of Life" (*Ibid*, 101–102), this is less the expression of mysticism than of taking "the fact of our being an animal species to be a fundamental feature of our existence" (Taylor 1981, 207). The "community" concept here is not one of mystical harmony, but of interacting species, each the product of evolution, subject to laws of genetics, natural selection, and adaptation. In other words, the concept of "life" in Taylor is derived from biology and ecology, not from a mysticism of "the will to live I find within myself."

Partly as a consequence of this, Taylor's ethic is much more objective than Schweitzer's. Schweitzer's approach to ethics is existential, with ethics as the answer to the question of how it is possible to live without despair while being true to the will to live. Schweitzer writes, "It is certain further that true ethics are always subjective, that they have an irrational enthusiasm as the very breath of their life" (Schweitzer 1923, 299). In contrast, Taylor argues that it is "rational" to accept the principle of respect for nature, and the claim to rationality does not have the existential origin Schweitzer demands. Taylor's grounding is much more objective than Schweitzer's. Not only does Taylor argue for a kind of disinterested rationality of accepting his principle of respect, but he also bases this rationality on an underlying belief system that he calls "the biocentric outlook on nature," which is not a scientific theory but rather "a philosophical world view" (Taylor 1981, 205). Thus, Taylor implicitly rejects Schweitzer's skeptical analysis of the project of basing ethics on a view of the world.

This has the further consequence that while Schweitzer writes, "In the history of ethics there is downright fear of what cannot be subjected to rules and regulations" (Schweitzer 1923, 291), Taylor thinks that acceptance of "a set of rules of duty" is the natural outcome of the attitude of respect for nature. Schweitzer's broad scope for subjectively authentic decision (under the overall guidance of reverence for life) is replaced by a set of rules that are, again, valid for every rational agent.

TAYLOR'S ETHICS OF RESPECT FOR NATURE

Just as "human ethics has to do with the moral relations holding among persons as persons" (Taylor 1986, 37), so environmental ethics is "concerned with the moral relations that hold between humans and the natural world" (*Ibid*, 3), where "the natural world" refers to ecosystems and the plants and animals that live in them, but excluding all "human intrusion and control" (*Ibid*, 3).[2] And where Schweitzer looks for a subjective and enthusiastic foundation for ethics, Taylor demands a *rationally grounded* system of moral principles (*Ibid*, 9). Taylor understands this requirement in a strong way. "We must strive for objectivity, and this requires a certain detachment from our immediate intuitions in this area so that we can consider without prejudice the merits of the case for a life-centered view" (*Ibid*, 24). This system of ethics should enable us to see that specific acts are either obligatory or forbidden. "[Q]uite independently of the duties we owe to our fellow humans, we are morally required to do or refrain from doing certain acts insofar as those acts bring benefit or harm to wild living things in the natural world" (*Ibid*, 10). The claim that the rules and standards of the system are *valid* means that "the rules and standards of the system are in truth binding upon (lay down requirements for) all moral agents" (*Ibid*, 25). The rules are to be applied "disinterestedly" in that they are "to be followed independently of the agent's particular ends and interests" (*Ibid*, 29).

One noteworthy aspect of Taylor's environmental ethic is that it is designed with an eye to establishing (or recognizing) a "symmetry" between human ethics and environmental ethics. Thus, human ethics "has to do with human beings as persons," where to be a person is to be "a center of autonomous choice and valuation" (*Ibid*, 33). The question human ethics has to answer is what it means to have an attitude of respect for persons as persons. Such an attitude would require one to "consider their personhood as having a worth or value in itself" (*Ibid*, 39), that is, to respect "each person's autonomy in living according to his or her chosen value-system" (*Ibid*, 38). Against this background, the theory has to involve specific duties. "The principle of respect for all persons as persons, then, imposes both requirements for the development of certain capacities in individuals and requirements for the free exercise of those capacities. The rules of a value system of human ethics must provide for the fulfillment of both requirements" (*Ibid*, 39–40).

Human ethics has three components (cf. *Ibid*, 41–42):

1. a *belief system* in which others are conceived of as persons like oneself;
2. the *attitude* of respect for persons;
3. a *system or rules and standards* that embody the attitude of respect for persons.

The symmetry between human ethics and environmental ethics requires that the latter have three components (*Ibid*, 44–47):

1. the *belief system* Taylor calls "the biocentric outlook on nature," in which one identifies oneself as a member of the biotic community (*Ibid,* 44), perceives each individual organism as "a teleological (goal-oriented) center of life" (*Ibid,* 45), and rejects the idea of human superiority (*Ibid,* 45);
2. the *attitude* of respect for nature, which requires that one "judge the good of each [member of a nonhuman species] to be worthy of being preserved and protected" (*Ibid*, 46);
3. a *system of rules and standards* that are morally binding on all moral agents. Of particular importance are rules that determine conditions under which the interests of one organism override the interests of another organism under conditions of conflicting interests (cf. *Ibid*, 192–198, 263–307).

Taylor devotes a chapter to each of these, beginning with the attitude of respect (chapter 2), then the biocentric outlook on nature (chapter 3), and finally rules and standards (chapter 4).

Thus, in *Respect for Nature*, Taylor begins with the second component, not with the first. (The essay "The Ethics of Respect for Nature" moves from the first to the second.) On the one hand, this ordering makes sense, because the biocentric outlook is introduced as that which "supports and makes intelligible" our commitment to the moral attitude (*Ibid*, 59). The investigation is regressive, moving from a discussion of an attitude we might adopt to a discussion of the reasons one might have for adopting it. Investigating the attitude in some detail first enables Taylor to *look for* the support and intelligibility required. But there are two dangers here. First, discussion of the attitude of respect prior to the discussion of the concept of the "community of life" might be conducive to naively taking one possible concept of "respect" (among others) as the concept. Second, taking one concept of "respect" as given, one might not notice that the concept of the "community of life" that supports it is itself in some way naïve, incomplete, or otherwise objectionable. This raises an interesting question: Does our concept of respect specify what a biocentric attitude has to be if it is to offer the required support? Or does the concept of the community of life play a role in determining what counts as the proper attitude of respect? Taylor makes a clear statement of the autonomy of ethics: moral conclusions cannot be deduced from the theories of the biological sciences (*Ibid*, 47). I will return to these questions below.

THE ATTITUDE OF RESPECT FOR NATURE

Taylor begins by investigating two crucial concepts. "In order to grasp what it means for rational, autonomous agents to take (or to have) the attitude of respect for nature as their own ultimate moral attitude, it is necessary to understand two concepts: the idea of the *good* of a being, and the idea of the *inherent worth* of a being" (*Ibid*, 59–60). While these two concepts are logically independent from one another, in morality they become connected in the following manner: *if* we adopt the attitude of respect for nature, then whenever we understand that a being has a good of its own we will also conceive it as having inherent worth. (Thus, to recognize a being as having inherent worth motivates the attitude of respect, and to respect a being because it has a good of its own is to take it to have inherent worth.)

A good of its own–The central concept in Tom Regan's animal rights approach is that of "taking an interest": any organism that *takes* an interest in the satisfaction or frustration of its own preferences is worthy of respect; any being that does not take an interest is not worthy of respect. To some critics, this means that the ethics of animal rights is in tension with environmental ethics, because trees and forests do not take an interest in their own well-being. Taylor has a more legitimate claim to developing an environmental ethic than does Regan, because Taylor focuses on the concept of "having a good of its own" rather than that of being the "subject of a life." To have a good of its own in this sense does not imply that the being in some sense *pursues* its good or takes an interest in it. Rather, to have a good of its own simply means that "it makes sense to speak of what is good or bad *for* the thing in question" and of "what does an entity good" (*Ibid*, 61). An entity has a good of its own if it makes sense to speak of something's being in that entity's interest. But that does not require that the entity *has* an interest in that thing, much less that it takes an interest in it.

Clearly, this concept of "having a good of its own" extends much more widely than does Regan's concept of being the "subject of a life," which involves having both preference and welfare interests (cf. Regan 1983, 214). For example, it applies to plants, since concepts like "survive in a healthy state," "develop in a normal manner," and "healthy adult" make sense when applied to plants (cf. *Ibid*, 66). In using these concepts, we "make value judgments from the perspective of the organism's life" (*Ibid*, 67).

Wherever it is possible to make such value judgments, "we can conceive of ourselves as having a duty to give consideration to its good and to see to it that it does not suffer harm as the result of our own conduct" (*Ibid*, 67), and this is the attitude of respect for nature. This is not to say that we *should* adopt such an attitude. It is merely an analysis of the concept of such an attitude. Taylor can claim to have shown only that the concept does not suffer from a hidden inner inconsistency (which would result if we wanted to recommend respect for trees using Regan's account of "respect").

Finally, given this account of what it means for an *organism* to have a good of its own, Taylor thinks that we can make sense of speaking of the good of a species-population (not a species itself[3]), or of an entire biotic community (*Ibid*, 68–71).

Inherent worth.[4]–Why should the fact that an entity has a good of its own make a moral claim on us? Put in that way, there is no such reason. Having a good of its own is a necessary but not yet sufficient condition of making a moral claim on us or of being a moral patient. This condition becomes sufficient only when we add the claim that the entity has *inherent worth*, since an entity that has inherent worth is worthy of respect (and only an entity with a good of its own can have inherent worth.) What, then, does it mean to say that an entity has inherent worth? To say that such an entity X has inherent worth is to assert the following:

> *A state of affairs in which the good of X is realized is better than an otherwise similar state of affairs in which it is not realized (or not realized to the same degree), (a) independently of X's being valued . . . by some human valuer, and (b) independently of X's being in fact useful in furthering the ends of a conscious being* . . . (*Ibid*, 75).

Note that this definition is strongly antiholistic—or, better, antirelational—in the sense that Taylor assumes that it makes sense to speak of State A in which X is m, and State B in which X is not m, as being "otherwise similar." Thus, consider the following example: The state of affairs in which a certain duck is not killed is better than the similar state of affairs in which the duck is killed.

In one sense, this is straightforward. The duck has inherent worth, so it is simply better (period) that it lives than that it dies. But this cannot be the whole story. The duck has inherent worth because it has a good of its own. So the "good (period)" is based on the fact that it is better *for the duck*. The question, "better for whom?" cannot be avoided, and once we ask this question it becomes clear that, when considering the good of the duck, the fox (or the human hunter) is being left out of the picture. (The underlying reasons for this will become apparent below.) Now the claim that the duck has inherent worth, while presupposing that the duck has a good of its own, seems to go further. By itself, the fact that the duck has a good of its own makes its escape good only for the duck. To say that the duck has inherent worth is to say that it is *bette*r (period) that the duck escape than that it die. But it is also true that it is *better for the* *fox and its young* that the fox succeed in killing and eating the duck AND that it is *better (period)* that the fox nourish itself and its kits than that they starve. Thus, it is better (period) that the fox kills the duck than that it and its kits starve. Put in Schweitzer's terms, Taylor's analysis is running into the self-alienation of the infinite will to live. I will take up this issue below.

Finally, the claim that an entity has inherent worth entails two moral judgments: "(1) that the entity is deserving of moral concern and consideration . . . and (2) that all moral agents have a *prima facie* duty to promote or preserve the entity's good as an end in itself and for the sake of the entity whose good it is" (*Ibid*, 75). Thus, once we recognize the inherent worth of an entity, it "is never to be treated as a mere means to human ends, since doing so would contradict—would amount to a denial of—its status as a bearer of inherent worth" (*Ibid*, 79).

How, then, can we recognize that an entity does or does not have inherent worth? This points to the second component of Taylor's theory, the biocentric outlook, since he argues that "all animals and plants in the natural world have inherent worth" (*Ibid*, 79–80), because "*only this way of regarding them is coherent with how we must understand them when we accept the belief-system of the biocentric outlook on nature*" (*Ibid*, 80).

THE BIOCENTRIC OUTLOOK ON NATURE

Taylor discusses the biocentric outlook on nature as that which "supports and makes intelligible" the attitude of respect for nature. "What moral significance the natural world has for us depends on the way we look at the whole system of nature and our role in it" (*Ibid*, 99). The biocentric outlook contains four components:

1. The belief that humans are members of the Earth's Community of Life in the same sense and on the same terms in which other living things are members of that community.
2. The belief that the human species, along with all other species, are integral elements in a system of interdependence such that the survival of each living thing, as well as its chances of faring well or poorly, is determined not only by the physical conditions of its environment but also by its relations to other living things.

3. The belief that all organisms are teleological centers of life in the sense that each is a unique individual pursuing its own good in its own way.
4. The belief that humans are not inherently superior to other living things (*Ibid*, 99–100).

Taylor claims that this attitude *will* be adopted by those who are "rational, factually informed, and have developed a high level of reality-awareness" (*Ibid*, 100).

Humans as Members of the Earth's Community

Human life is "*an integral part of the natural order of the Earth's biosphere*" (*Ibid*, 101; Taylor's emphasis), and in this sense our place in nature is the same as that of any other species, since we, like them, are biological. "Full awareness of this common relationship gives us a sense of true community with them" (*Ibid*, 101).[5]

What is this "sense of true community"? We share a common origin in evolution, and we share the natural environment with other species; we are not something "set apart." "We are then ready to affirm our *fellowship* with them as *equal members* of the whole Community of Life on Earth" (*Ibid*, 101; emphasis added). The italicized words have to be examined very carefully, since they are loaded with moral connotations from their role in human ethics.

Taylor roots the specific sense of "membership" at work here in five realities (*Ibid*, 101–102):

(a) shared biological and physical requirements for survival and well-being,
(b) the fact that all have "a good of their own,"
(c) a shared "sense of freedom,"
(d) the fact that we humans are a recent arrival on the stage of life,
(e) the fact that species need each other (this leads to the next component).

"Awareness and acknowledgment" of these realities "constitute the basis for our conceiving of ourselves to be but one small part of the total membership of the Earth's Community of Life" (*Ibid*, 102).

Concerning (a), the lesson is that our own preservation and flourishing require that we orient ourselves to biological requirements as "our normative guides" (*Ibid*, 103). (This has the structure of a hypothetical imperative: IF we want to survive and flourish, THEN we must ensure that the biological prerequisites of this are guaranteed.) We are members of the earth's community in the sense of biological dependence.

Concerning (b), not only are we not the only beings with a good of our own, including survival and well-being, none of us have total control of the biological and physical requirements mentioned under (a). In becoming aware this, "we come to see that we are in the *same existential situation* as are all other living things" (*Ibid*, 105; emphasis added). We are members of the earth's community in the sense of sharing a common fate.

Concerning (c), this is not a denial of the uniqueness of human free will, autonomy, and social freedom. But there is an additional, shared freedom in the sense of "being in a position to be able to preserve one's existence and further one's good," where "freedom" means "absence of constraint" (*Ibid*, 106). "An organism may be said to be free if it has the ability and opportunity to promote or protect its good according to the laws of its nature" (*Ibid*, 109). A caged animal is not free in this sense, and we speak of *freeing* an animal in such situations. Taylor concludes that we "are united with other forms of life under this concept of freedom," and that "we constitute a community of beings in part because we have this value in common" (*Ibid*, 111). We are members of the earth's community in the sense of sharing a need for freedom.

Concerning (d), while we share a common origin with all forms of life, we are a recent arrival on the stage of evolution. As a result, "we understand ourselves as beings that fit into the same structure of reality that accounts for every other form of life," and this constitutes "a vivid sense of our membership in the Earth's total Community of Life" (*Ibid*, 113). From this perspective, it makes no sense to say that we are the final goal and culmination of the evolutionary process. We are members of the earth's community in the sense that we are newcomers and not the telos of the process.

Concerning (e), we are dependent on the earth's biosphere, but it is not dependent on us. We are "needy dependents," whereas our disappearance would actually be beneficial to many species. We are members of the earth's community in the sense that we need it while it does not need us. (This is largely a dramatic restatement of (a).)

As a result of these considerations, the "sense of oneness," of seeing "ourselves as members of one great Community of Life" (*Ibid*, 115), is partly a matter of identifying with other forms of life because we have a sense of being in the same boat (a–c) and partly a matter of having a sense of being precarious newcomers (d–e).

THE NATURAL WORLD AS A SYSTEM OF INTERDEPENDENCE

This component of the biocentric outlook develops several motifs that contribute to the sense of membership. We are members of the earth's community in that like all life we face biological and physical requirements for survival and well-being, but in addition to this we are interdependent in the sense that life is "a tightly woven web" (*Ibid*, 116).

Taylor's example is that of alligators in the ecosystem of the Florida Everglades. Alligator nests hold water during the dry season, and contain various forms of life that are nourished by alligator droppings. These nests become the core of islands, which support trees, which in turn support nesting birds, which in turn live on fish and insects, "maintaining the balance between animal and plant life in the grasslands" (*Ibid*, 116). When humans trap alligators, the entire ecosystem suffers, endangering the system as a whole (cf. *Ibid*, 116–117). The emphasis in the description is on symbiosis. Predation is mentioned twice (birds on fish and insects, alligators on nesting birds), each time balanced by the positive contribution made to the system by the predation. Human trapping is outside interference that destabilizes and endangers the entire system. The lesson is one of complex interdependence: a radical change in one element of the system causes an adjustment in others. "It is for this reason that the entire biosphere to our planet comprises a single unified whole, which I refer to in this book by the term 'the natural world'" (*Ibid, 117*).

Thus, if we see ourselves as members of the earth's community, "We will see ourselves as an integral part of the system of nature" (*Ibid*, 117). The difference between us and other members of the community is that we have to decide which role we are to play. We cannot opt out, but our awareness of interconnection (and of the fact that even as our knowledge of the structure of these interdependencies grows, the biological conditions required for our survival and well-being are never totally under our control) can have an enormous impact on the way we choose to live.

Taylor is careful to point out that this emphasis on interdependence in the biocentric outlook does not mean that he is committed to a holistic or organicist view of environmental ethics. In other words, "We cannot derive moral rules for our treatment of the natural world from a conception of the Earth's biosphere as a kind of supraorganism, the furtherance of whose well-being determines the ultimate principle of right and wrong" (*Ibid*, 118). The health or good of the system as a whole is of moral significance only because the individuals that make up the system themselves have a good of their own that is deserving of moral consideration. "Once we have rejected a human-centered ethics, it is only by reference to the particular lives of such beings as made better or worse by our actions that consideration for the natural world becomes morally relevant" (*Ibid*, 119). Taylor's argument here is that the health of the system as a whole is of instrumental value to the individuals that make up the system, but it is only those individuals who have inherent value, not the system as a whole.[6] It follows that it makes little sense to speak of having "respect" for the system as a whole, except in the indirect manner indicated.

Organisms as Teleological Centers of Life

While the focus on the system of interdependence might seem to reduce each individual organism to its function within the whole, Taylor thinks that this would destroy environmental ethics.[7] This tendency and attendant danger is counteracted by science's increasing ability to "enable us to grasp the uniqueness of each organism as an individual" (*Ibid*, 120). When one pursues this line of knowledge, one ultimately "achieves a genuine understanding of its point of view" (*Ibid*, 120). Such an ability to conceive of an organism as a teleological center of life need not involve a false anthropomorphizing. Indeed, one need not assume that the organism in question takes an interest in its good at all. The fact remains that both trees and protozoa "have a good of their own around which their behavior is organized" (*Ibid*, 122). In this sense, each individual living thing "is seen to have a single, unique point of view" (*Ibid*, 122), whereas a stone has no point of view.[8]

Our awareness of individual organisms as teleological centers of life has, in principle, the features of *objectivity* and *wholeness of vision* (cf. *Ibid*, 125–127). Such a view is objective in the sense that the aim is to understand the organism as it is, not as we want it to be (as evil, good, innocent, etc.) (*Ibid*, 125–127). The goal is wholeness of vision in that the focus is on the animal itself, not on some function it might play in our lives. Taylor's (negative) example here is the hunter, who views the pheasant as something to be shot during game season, the bird's life outside of hunting season being of no interest to "him" unless it somehow relates back to the status of game bird (*Ibid*, 127).[9] This objectivity and wholeness

> *sets a relevant frame of reference for our conduct as moral agents. As a result of our heightened awareness of the reality of another living thing's existence, we gain the genuine capacity to take its standpoint and make judgments based on its good. Shifting out of the usual boundaries of anthropocentricity, the world-horizon of our moral imagination opens up to encompass all living things. Seeing them as we see ourselves, we are ready to place the same value on their existence as we do on our own. If we regard ourselves as possessing inherent worth, so will we be disposed to regard them* (*Ibid*, 128).

Note that while exercising the capacity to take the standpoint of an organism is a necessary condition for the attitude of respect for nature, it is not sufficient. It makes the attitude of respect possible, but "having this capacity does not necessitate our making the moral commitment" (*Ibid*, 129). That depends on the next element of the biocentric outlook.

The Denial of Human Superiority

Most of the Western cultural, religious, and philosophical traditions are committed to a value hierarchy in which human beings stand at the pinnacle of creation or evolution. Often the conclusion is explicitly drawn that human existence is valuable in itself, while animal life is there for human use and thus has only instrumental value. This justifies human beings in using animals as they see fit, with the occasional warning against cruelty (and even that is often based on the claim that cruelty to animals hardens our hearts and makes us more likely to be cruel to human beings, only the latter being directly immoral). Here Taylor launches an attack on the prejudice of human superiority that is similar to that found in Regan's work. He argues that claims to human superiority are made "from a strictly human point of view" (*Ibid*, 130), and this makes arguments for human superiority viciously circular: they are based on the presupposition of the superiority of the human point of view in order to demonstrate human superiority.

Once this circularity is seen, one must face the possibility that human life and other forms of life are of *equal* inherent worth.

> *A total reordering of our moral universe would take place. Our duties with respect to the "world" of nature would be seen as making claims upon us to be balanced against our duties with respect to the "world" of human culture and civilization . . .*
>
> *Now, this radical change in our view of the natural world and of our proper ethical relationship to other living things is just what is involved when we accept the biocentric outlook. The fourth component of that outlook is nothing else but the denial of the doctrine of inherent human superiority. This denial, as we shall see, is the key to understanding why acceptance of the biocentric outlook underlies and supports a person's adopting the attitude of respect for nature* (*Ibid*, 134).

Acceptance of the first three elements of the biocentric outlook—viewing humans as members of the Earth's Community of Life, viewing nature as a system of interdependence, and viewing individual organisms as teleological centers of life—prepares the way for the acceptance of equal inherent worth (cf. *Ibid*, 153–54) and of the principle of "species-impartiality" (*Ibid*, 155). Finally, to regard organisms as having equal inherent worth is to consider them with the attitude of respect. With the denial of human superiority, the biocentric outlook yields the attitude of respect for nature (*Ibid*, 155).

The biocentric outlook is thus a world view in Schweitzer's sense, one having several salient features.

- It shows us that we are *biological creatures*, and in this sense we are no different from other living things. Our biological "oneness" with other living beings is acknowledged.
- It shows us that we are "an integral part, along with every species that shares the Earth with us, of a world order that is

structured in a certain way" (*Ibid*, 157). We learn that just like other living creatures we are "functionally interdependent units" of broader biotic communities.

- Our awareness of the point of view of other living creatures as teleological centers of life is heightened.
- Finally, given these three aspects, we tend to view all animals and plants "in an impartial light" (*Ibid*, 157).

But why should we adopt this outlook? Taylor argues first that this outlook is *acceptable* in the sense that it satisfies traditional criteria such as comprehensiveness and completeness, systematic order, freedom from obscurity, and consistency with known empirical truths (*Ibid*, 158–165). Given this, he goes on to argue that "insofar as moral agents are rational, factually enlightened, and have a developed capacity of reality-awareness, they will adopt those criteria as the basis for deciding what outlook on nature to accept as their own" (*Ibid*, 158) and thus will adopt the biocentric outlook (cf. *Ibid*, 165–168). And it follows that the closer we approximate such "ideally competent evaluators" (*Ibid*, 165), the more we will be disposed to adopt the biocentric outlook.

Note that while the biocentric outlook is not identical with the view of the world provided by the science of ecology, the criteria for acceptability Taylor offers are derived from the criteria for the acceptability of scientific theories. Consistent with this, the characteristics of the ideally competent evaluator of an outlook include *objectivity* and *detachment* (*Ibid*, 162).

THE SYSTEM OF RULES AND STANDARDS

My discussion of Taylor's system of rules and standards is brief, because I am concerned only with the general idea of such rules and standards. I first sketch the general nature of the principles, and then examine the rules and standards specifically concerning hunting.

A system of rules is required because the attitude of respect is an "ultimate" attitude (cf. *Ibid*, 41–42 concerning respect for persons, 90–98 concerning respect for nature). The first problem this raises is one of justification. Since the attitude of respect is in each case ultimate, it cannot be derived from a more basic attitude (and is thus not open to "proof"). But when it comes to application of the principle of respect, other problems arise, especially in the case of respect for nature. For Taylor, respect for nature involves respect for individual living creatures as teleological centers of life. For each such creature, respect gives rise to the "*prima facie* duty to promote or preserve the entity's good as an end in itself" (*Ibid*, 75). But of course the fact that the lives of individual living creatures are interdependent (a central claim in Taylor's biocentric outlook) means that promoting or preserving the good of one entity will in fact be at the cost of another entity's good.

A comparison with Schweitzer is instructive at this point. Schweitzer's principle of reverence for life involves a commitment to life as such, in abstraction from the intertwining of life and death. His mysticism of the infinite will to life that underlies all individual lives enables him to make this distinction. But when one attempts to realize reverence for life in this world in which life is divided against itself, one discovers that the principle inevitably gives rise to situations of moral conflict. For Schweitzer, this does not indicate a fundamental flaw in his approach to ethics, since he explicitly set out to ground his ethical view of the world in his view of life, which is in turned based on the will to live as it manifests itself in individuals. But Taylor's situation is different. Taylor bases his ethical view of life on the biocentric outlook as a philosophical view of the world. It might therefore seem more disturbing from a systematic point of view to discover that the object of respect is, as it were, "divided against itself." Given the fact that the recognition of interdependence is an essential part of the biocentric view, one might expect that this division of life against itself would already be built into the principle of respect, as one aspect of that which we are to respect. But it is not, and this gives rise to the problem of providing priority principles. This is another point of dissimilarity with Schweitzer, who thinks that fundamental reality of life divided against itself forces us to make subjective decisions of personal authenticity within the framework of reverence for life. Schweitzer vehemently opposes a "relative" ethics that attempts to mediate the conflicts, playing *prima facie* duty off against conflicting *prima facie* duty to arrive at a concrete duty that can be performed without guilt. Taylor is much more traditional. He wants objective rules of duty and objective priority principles. There is much less room for decisions that are both subjective/individual and moral (cf. the discussion of the ethical ideal of harmony below).

Priority principles are required in three types of situation: situations in which our duties to nature conflict with one anther, situations in which our duty to nature conflicts with our duty to human beings, and situations in which our duties to human beings conflict with one another. In *Respect for Nature* Taylor is concerned only with the first two. Rules of duty in environmental ethics include:

- *The rule of nonmaleficence* (*Ibid*, 172–173),
- *The rule of noninterference* (*Ibid*, 173–179), which includes duties not to place restrictions on individual organisms and the duty to maintain a "hands off" policy toward ecosystems and biotic communities,
- *The rule of fidelity* (*Ibid*, 179–186),
- *The rule of restitutive justice* (*Ibid*, 186–192).

These rules make sense primarily for a moral agent that is as a matter of empirical fact part of (and thus able to have an effect on), yet not (as such) essentially a participant in the natural order. They can, of course, conflict with one another in a concrete situation, and Taylor works out a set of priority rules for such situations (*Ibid*, 192–198).

More importantly, such rules can conflict with our duties to human beings, and Taylor devotes the last chapter of *Respect for Nature* to this problem (*Ibid*, 256–314). He develops five priority principles for such situations:

- *The principle of self-defense* (*Ibid*, 264–269);
- *The principle of proportionality* (*Ibid*, 269–280), which depends on the distinction between basic and nonbasic interests;
- *The principle of minimum wrong* (*Ibid*, 280–291);
- *The principle of distributive justice* (*Ibid*, 291–304);
- *The principle of restitutive justice* (*Ibid*, 304–307).

Exploitation

In light of these duties, Taylor condemns a series of practices as being "exploitative." "[A]ll such practices treat wild creatures as mere instruments to human ends, thus denying their inherent worth. Wild animals and plants are being valued only as a source of human pleasure or as things that can be manipulated and used to bring about human pleasure" (*Ibid*, 274). Taylor's list of examples is interesting:

- Slaughtering elephants so the ivory of their tusks can be used to carve items for the *tourist* trade.
- Killing rhinoceros so that their horns can be used as *dagger handles*.
- Picking rare wildflowers, such as orchids and cactuses, for one's *private* collection.
- Capturing tropical birds, for sale as caged *pets*.
- Trapping and killing reptiles . . . to be used in making *expensive* shoes, handbags, and other "*fashion*" products.
- Hunting and killing of rare wild mammals, such as leopards and jaguars, for the *luxury* fur trade.
- All hunting and fishing done as an *enjoyable* pastime (whether or not the animals killed are eaten), when such activities are not necessary to meet the basic interests of humans. This includes all *sport* hunting and *recreational* fishing (*Ibid*; 274, emphasis added).

The italicized words point to the nonbasic human interests that drive such activities. (The two appearances of the word "rare" might seem to indicate that collecting orchids or killing wild animals for the luxury fur trade might be permissible as long as they are not rare, but that is clearly inconsistent with Taylor's principles.) To call such practices "exploitative" indicates that they are inconsistent with recognition of the inherent worth of the animals and plants involved. Another way to put this is that these practices are not justified by any priority principles that could override our respect for and *prima facie* duties to the animals and plants.

Hunting

The titles of these duties make it clear that, for Taylor, hunting wild animals is *prima facie* immoral.

The intention to kill is part and parcel of hunting, so it is essentially maleficent in that sense. In Taylor's example, if a wild falcon kills a mouse, nothing immoral has occurred, since the falcon is not a moral agent, though something bad, inherently bad, has occurred. When a mouse is killed by a Peregrine falcon being hunted under the direction of a falconer, something immoral has occurred—on the part of the falconer, not the falcon (Taylor 1986, 172–173). (This example is in complete agreement with Schweitzer's condemnation of falconry. Cf. Schweitzer 1950, 174–179).[10]

Hunting interferes both with the lives of individual animals, restricting or destroying their "natural freedom" (Taylor 1986, 175), and with the functioning of whole ecosystems. Hunting, or even transplanting wild plants to a garden, violates "the duty to let wild creatures live out their lives in freedom" (*Ibid*, 174). Respect for nature requires that we not "intrude into the domain of the natural world and terminate an organism's existence as a wild creature" (*Ibid*, 175). "[A]s far as our proper role as moral agents is concerned, we must keep 'hands off.' By strictly adhering to the Rule of Noninterference, our conduct manifests a profound regard for the integrity of the system of nature" (*Ibid*, 176). (Recall here Taylor's example of alligators in the Everglades (*Ibid*, 116–117).) It is clear that this rule is not the expression of an essential involvement of human beings qua moral agents in the natural world. "This general policy of nonintervention is a matter of disinterested principle" (*Ibid*, 177). We are, of course, involved in and interested in the natural world as a matter of fact, but this contingent fact does not change the moral principle. We may want to help a wild animal, but the morality of respect for nature forbids us to do so (*Ibid*, 177). For Taylor, this principle is an expression of our recognition that "suffering and death are integral aspects of the order of nature" (*Ibid*, 177), and that this is quite in order as long as that suffering and death are natural. Contra Schweitzer, "nothing goes wrong in nature" (*Ibid*, 177), nature freed from human intervention, that is.

Hunting essentially involves stealth, deception, and betrayal of trust (cf. *Ibid*, 180–182). "Deception with intent to harm is of the essence" of hunting, trapping, and fishing (*Ibid*, 179). It thus manifests a lack of respect for the animals involved. And while hunters may proclaim their "respect" for the animal they hunt, for Taylor it is clear that such respect is respect only for some excellence of the individual animal (large antlers) or of the species (wariness), and not respect for the animals as having a good of their own and thus having inherent worth. Using a distinction analyzed by Stephen Darwall, Taylor writes, "It should be clear, however, that in all such claims to respect the animal it is *appraisal* respect, not recognition respect, that is meant" (*Ibid*, 183). One may respect the wariness of deer, the fighting spirit of bass, and so on (appraisal respect), but this is not respect based on the intrinsic worth of the animals (recognition respect).

These duties establish that hunting is *prima facie* immoral. They may be overridden by the requirements of restitutive justice, and in subsistence hunting our duties to animals may be overridden by our duties to human beings. But the ban on "sport hunting and recreational fishing" is absolute (*Ibid*, 274). In subsistence hunting "such activities are not done as enjoyable pastimes but out of necessity" (*Ibid*, 275). This still leaves the problem of the hunter who hunts out of necessity AND finds hunting a deeply meaningful, fulfilling, and even enjoyable activity. I think that Taylor would have to say that the attitude, if not the very activity, is immoral. Just as the theory of respect for nature needs a theory of virtue (*Ibid*, 198–218), it also needs a theory of vices. Enjoyment of hunting, even when the hunting itself is necessary, would be such a vice.

THE UPSHOT: THE ETHICAL IDEAL OF HARMONY

Taylor acknowledges that even this set of rules and principles is not sufficient to resolve every case of conflict. Yet he thinks that our decisions need not be arbitrary. "We can avoid arbitrariness and randomness in these situations, I suggest, by referring to our total picture or vision of what kind of world order would be ideal according to the structure of normative principles we have accepted" (*Ibid*, 307). The point of the ideal is to present us with an image of the world as it would be if all moral agents always manifested respect both for persons and for nature. "The most apt phrase for describing this 'best possible world' in its simplest terms is: *a world order on our planet where human civilization is brought into harmony with nature*" (*Ibid*, 308). What, then, is involved in this kind of harmony?

As it is used here "harmony" means the preserving of a balance between human values and the well-being of animals and plants in natural ecosystems. It is a condition on Earth in which people are able to pursue their individual interests and the cultural ways of life they have adopted while at the same time allowing many biotic communities in a great variety of natural ecosystems to carry on their existence without interference. Whatever harm comes to the individual members of those communities results from the ongoing processes of evolution, adaptation, and natural changes in environmental circumstances, not from human actions (*Ibid*, 309).

Chapter 5

Critical Analysis of Taylor's Philosophy of Respect for Nature

ENVIRONMENTAL STOICISM: THE IDEAL OF HARMONY

I begin my critical discussion of Taylor's philosophy of respect for nature by looking more closely at its final element, the ideal of harmony. My goal in this chapter will be to trace an implication I find in Taylor's ideal back to its root in his use of the biocentric outlook on nature as that which "supports and makes intelligible" a commitment to the moral attitude of respect for nature as he understands it (*Ibid*, 59).

When we take a second look at Taylor's sketch of his vision of a "best possible world," it becomes apparent that in such an ideal world *human beings would not be members of the Earth's Community of Life* or, more to the point, *human beings would not be part of the natural world as a system of interdependence.*[1] This is what I refer to as Taylor's environmental stoicism: Taylor's ideal of human disengagement from the natural world is a loose parallel to the stoic ideal of *apathia*, of freeing oneself from the passions and affections.[2] Only under this condition is it even thinkable that the only harm to individual living beings would be the result of *natural* (meaning non-human) *forces*. Taylor tacitly admits this when he writes,

> *In this ethical ideal our role as moral agents is to direct and control our conduct so that, with regard to animals and plants living in the wild, we comply with the four basic rules of environmental ethics* . . . Although we cannot avoid some disruption of the natural world *when we pursue our cultural and individual values, we nevertheless constantly place constraints on ourselves so as to cause*

> *the least possible interference in natural ecosystems and their biota. The realm of nature is not considered as something to be consumed, exploited, or controlled only for humans [sic] ends, but is shared with other creatures* (*Ibid*, 309–310; my emphasis).

Taylor acknowledges, of course, that we cannot escape interdependence, the recognition of which plays an important role in the biocentric outlook on life. Indeed, he goes to some lengths to make things as "harmonious" as possible, given this interdependence. Thus, the Earth is to be divided up into areas for humans and areas for wild animals (although it remains unclear how the wild animals such as deer are to be kept out of areas devoted to agriculture), and where this is impossible, humans and nonhumans are to "take turns" in using habitat (*Ibid*, 310). In unpublished notes on human relations to what he calls the "bioculture"—in this case, to domesticated plants—Taylor makes a distinction between eating that which can be removed from a plant without killing it (nuts, seeds, fruits, etc.) and eating entire plants (lettuce, carrots, etc.), which involves killing the plants. Soybeans are viewed as a legitimate source of protein, since "soybeans can be harvested just after the plants die each year" (Taylor 1990; cf. the report in Hettinger, 12, ftn. 21). Taylor is clear that even if we were to follow all of his rules very carefully, the result would not amount to the idealized harmony, but it remains the ideal that offers us guidance in making our decisions and in this way allows us to avoid arbitrariness.

This ideal is not consistent with two central components of Taylor's biocentric outlook on nature. Contrary to the biocentric outlook's official proclamation that human beings are integral member of the Community of Life "in the same sense and on the same terms in which other living things are members of that Community" (Taylor 1986, 99) and that this Community is a system of interdependence, the attitude of respect for nature, as Taylor understands it, separates human beings from nature and subjects them to the *prima facie* duty to stay out of nature. Of course, this removal cannot be maintained, and that is what the priority principles are designed to take care of. But from the point of view of a truly biocentric outlook, one that takes interdependence seriously, the priority principles are a series of *ad hoc* attempts to deal with the inevitable consequences of the view that human beings *should not*, from the moral point of view, be part of nature[3] while recognizing that, as a matter of empirical fact, human beings *are* part of nature.

There is, in other words, an unresolved tension between Taylor's biocentric outlook and the attitude of respect that it supports. One aspect of this outlook is that human beings are viewed as an integral part of the Community of Life as a system of interdependence, while for Taylor the end result of taking the biocentric outlook seriously is that we are to remove ourselves as completely as possible from wild nature, the ideal being literally zero impact. Human beings are both part of nature—and need to be more conscious of this fact than they have been in the past—and not alienated enough from nature. A closer look at Taylor's notion of "respect for nature" shows how Taylor arrives at this impasse.

THE CENTRALITY OF INTERDEPENDENCE

The tension I find in Taylor's ideal of harmony has its roots in his biocentric outlook on nature and in the concept of "respect" that is based on it. I therefore first examine his characterization of the biocentric outlook on nature and then turn to his concept of "respect." I thus reverse the order of discussion Taylor chooses in his book. Recall that there are four component beliefs of the biocentric outlook as Taylor sees it:

1. like all living things, human beings are members of the Earth's Community of Life;
2. like all other species, human beings are caught up in a system of interdependence;
3. organisms are teleological centers of life;
4. humans are not inherently superior to other living things.

When we look at the individual components of the biocentric outlook on nature and ask what role they play in providing support for the attitude of respect for nature, it is clear that the central role is played by 3—the belief that all living things are teleological centers of life. This is the most crucial component because being a teleological center of life means that the organism has a good of its own, which in turn means that the organism has inherent worth. By itself, this would mean that organisms are deserving of moral consideration, but it would also leave the door open to a hierarchical conception of inherent worth such that human beings are inherently superior to nonhuman organisms. Thus, 4, the belief that humans are not inherently superior to other living things, also plays a crucial role, since the denial of human superiority means that the interests of all organisms are deserving of equal consideration.[4] The natural result is Taylor's attitude of respect for nature, which involves the *prima facie* duty to protect and preserve all living things for their own sake.

Component 1, the belief that all living things are members of the one Community of Life, is used to provide emotional support to the attitude of respect, since it is to provide a felt sense of "true community" and "fellowship" with wild animals and plants. Taylor's use of capitalization in the phrase "the Earth's Community of Life" is both indicative of this felt sense of kinship and fellowship and quietly reinforces it. But it should also be noted that this does not fit very comfortably with Taylor's call for objectivity and "a certain detachment from our immediate intuitions in this area so that we can consider without prejudice the merits of the case for a life-centered view" (*Ibid*, 24).

Component 2, the belief that all living things are part of an interdependent web of life, has not even been mentioned. And indeed, it is of no *direct* relevance for the either the concept of the Community of Life or the attitude of respect *as Taylor develops it*. The only direct lesson Taylor derives from it in his discussion of the biocentric outlook on nature is the hypothetical imperative that if we are concerned with our own survival and well-being, we need to pay attention to the well-being of the web of interdependence on which we, like all living things, depend.

The belief that organisms are teleological centers of life is the key to Taylor's moral focus on individual organisms (his individualism) as opposed to a focus on systems such as the biosphere or ecosystems (the holism Taylor rejects in *Respect for Nature*). These systems are the systems of interdependence of 2, which are thereby marginalized from the moral point of view. There is nothing hidden or surreptitious about this. Taylor insists that environmental ethics cannot be holistic or organicist, since unless individuals have a good deserving of moral consideration, there is no reason why the system should be preserved (cf. *Ibid*, 118–119). Taylor thus essentially poses an either/or: either we are holists who have only a functionalist, and thus instrumental, regard for the good of individuals and no account of the moral status of the whole, or we are individualists who recognize only an indirect value of systems. I shall return to this dilemma below.

But 2 also has another role to play in Taylor's environmental ethic. It is the inescapable fact of interdependence that makes conflicts of interest inevitable, and Taylor develops the priority principles to guide our actions in situations of conflict of interest. So while the fact of interdependence does not play a direct role in either forming or supporting the attitude of respect for nature, it does play an important role in Taylor's full-blown ethics of respect for nature.

INTERDEPENDENCE AND RESPECT

I want to ask what effect it would have on the attitude of respect if the fact of interdependence were taken as directly supporting that attitude, and thus playing a role in determining the proper *sense* of "respect." I am going to suggest that it would have a powerful effect. But more than that, I want to argue that this shift in emphasis among the components of the biocentric outlook on nature is not optional. Rather, it is unavoidable, under threat of the incoherence of the entire position. In other words, one cannot first adopt Taylor's attitude of respect for nature, and then selectively draw on the components of the biocentric outlook in such a way that they provide support. The biocentric outlook can *only* provide support for an attitude of respect that is very different from Taylor's.

My point is that the fact that living things are teleological centers of life cannot be separated, even in thought, from the fact that all life is caught up in a system of interdependence. For an organism to be a teleological center of life, for it to have a good of its own, *presupposes* that it relates to its environment instrumentally as it pursues its own good, and for all animal life that means using other teleological centers of life. Since for Taylor "inherent worth" is based on an organism's having a good of its own, inherent worth is inextricably bound up with instrumental value. In short: no inherent worth (a being that has or pursues its own good) without instrumental value (that which is appropriated in that pursuit); and *vice versa*: no instrumental value (something is appropriated by another being) without the inherent worth of the being that uses that thing instrumentally.

If we take up the point of view of an animal or plant as a teleological center of life, as Taylor would have us do, we take up that perspective as one center among others in a complex web of interdependence. To take up one particular perspective is to view the surrounding world of that creature as a set of opportunities (and threats) that are structured in terms of that center. To return to the earlier example, from the point of view of the fox, it is good (for the fox) that the fox kills the duck. The fact that it *can* be good for the fox is the very source of the fox's inherent worth, and one cannot leave the duck out of what is good for the fox without leaving out the fox's good itself. By the same token, to say that the state of affairs in which the good of the duck is realized is better than an "otherwise similar" state of affairs in which the good of the duck is not realized (this was the formula of inherent worth) simply does not make sense, since in leaving the fox out, interdependence is left out, and interdependence is a fundamental and biologically constitutive fact about what it means to be a teleological center of life.[5] In line with this criticism, note that saying that: It is better (period) that the duck escape than that it die, and It is better (period) that the fox nourish itself and its kits than that they starve, is contradictory when one puts them in their properly relational form: It is better (period) that this duck escape this fox at this moment than that the duck die, and It is better (period) that this fox kill this duck at this moment than that the fox and its kits starve.

Talk of the equal inherent worth of the fox and the duck is fine—indeed true—but the comparatives point to a deep incoherence in Taylor's account.

This means that when the basic principle of environmental ethics, the principle of respect for nature, is taken to mean not only that one is to "judge the good of each [member of a non-human species] to be worthy of being preserved and protected as an end in itself and for the sake of the being whose good it is" (*Ibid*, 46), but in addition, "that all moral agents have a *prima facie* duty to promote or preserve the entity's good as an end in itself and for the sake of the entity whose good it is" (*Ibid*, 75), this *prima facie* duty in effect requires that moral agents view themselves as being, qua moral agents, free from or external to the system of interdependence.[6] In other words, the belief that we are "members of the Earth's Community of Life," and with it the sense of "true community" and "fellowship" to which it gives rise, are being decoupled from the fact that this community is a system of interdependence in which organisms, all organisms, live by making use of other things, generally other living things.

I want to suggest that recognition of this problem in Taylor's philosophy points to the path between the Scylla of unbridled anthropocentrism—in which the very meaning of the existence of the nonhuman world is found in its being useful to human beings—and the Charybdis of a biocentrism whose ideal is to remove us qua moral agents from nature, only to bow to reality by introducing principles that qualify the basic principle of environmental ethics, the principle of respect. My objection is not that a given position might not go far enough, or go too far, in allowing human interests to override the interests of other living things. The objection is rather that some special permission, strong enough to override the *prima facie* duty not to interfere, is required for human biological appropriation in general, of any kind, to be permissible for a moral agent. In other words, environmental ethics needs to recognize that respect for nature does not require that *prima facie* we ought not to interact with wild nature. What it should require is that, as we *participate* in the natural world, we do so *respectfully*. This requires a very different concept of "respect" than Taylor's, since such a concept must not only integrate the fact of interdependence, it must express an *affirmation* of the goodness of interdependence. As I will argue in Part IV, not all instrumental use of living things involves treating them as *mere* instruments.

Put in different terms, just as the inherent worth of any living being is inextricably bound up with the instrumental value of that being's environment, so the concepts of inherent worth and instrumental value are not and cannot be simple moral contraries: to recognize and respect the inherent worth of something does not commit one *ceteris paribus* to leaving it alone, or protecting and preserving it. In this light, it is possible that recognition of the inherent worth of wild animals *as such*, respect for their very wildness, can both require of us that we protect the threatened habitat and reduce human impact on it to the level that allows the animals to thrive, *and* allow or even call us to participate in that wildness, in that inherent value of the wild animal, for example by hunting it—hunting it in a way that respects and affirms both its wildness and our own. Once interdependence is acknowledged *and affirmed*, hunting is not contrary to respect, but can rather be an expression of it. The Kantian injunction not to use any *person* as a mere means has its counterpart here—but the counterpart is not an *extension* of the Kantian principle, which, as I have argued above, has its proper sense in human ethics and there alone. The proper sense of respect can only be determined in the context of that which is to be respected, which is to say in the context of our experience of and interaction with that which we strive to respect.

INTERDEPENDENCE AS MEANING—PAUL SHEPARD

What, then, is the alternative? For purposes of exploration, I consider the way suggested by Paul Shepard in an essay entitled "To Hunt." Shepard's formulation of the problem sounds like Schweitzer: "Causing death—requiring it in order to live—is the blade pricking the throat of human conscience" (Shepard 1991, 45). But his reply contains an implicit rejection both of Schweitzer's abstract mysticism of life[7] and of Taylor's disinterested ethics of respect. Noting the apparent parallel between human liberation movements and the "animal liberation" movement of the 1970s, Shepard argues that "the moral rejection of racism within our own species cannot be extended to a rejection of death-dealing between species. This would thwart an essential factor of organic existence. Diversity and kinship of life not only include the fact of death but require it" (*Ibid*, 45). Shepard is suggesting what might be called an *ecological outlook on the world*, one in which something like Taylor's concept of a system of interdependence is taken as its point of departure.[8] This is not an outlook on nature as opposed to persons, but on the world of which persons are a part. Rather than taking the science of ecology as the model for environmental ethics (as Taylor accuses Aldo Leopold of doing), Shepard poses an *existential* question: what is the proper *meaning* of life lived at the expense of life? And rather than appealing to a detached, scientific appreciation of individuality, Shepard directs us to an examination of human lives that could not hide the necessity and reality of death in slaughterhouses or cloak it in the claimed moral superiority of what Ted Kerasote calls "fossil fuel vegetarianism" (Kerasote 1994, 185, 232–236).

> *The earliest human art and ritual celebrate the epiphanies of hunting great game animals. They tell of the quest that grew up around the killing of animals—a kind of double seeking,*

> *both physical and spiritual. Those early artists were familiar with the living stream, and at home on earth they sought an accommodation of meaning as they hunted their food—and sometimes died themselves as food for other species . . .*
>
> *Where we see the consumption of animals as participation in the transfer of energy, as do traditional hunters, they also see it as the movement of an endless spiritual flow. To live and die is to be surrounded by beings whose coming and going are intrinsic to life itself (Shepard 1991, 45).*

When the problem is posed in this manner, as a problem of meaning within the overall flow of life, one of the traditional tools of animal rights/liberation/respect thought is radically reframed: the idea of using animals "merely as a means." It is not that advocates of this approach are blind to the phenomenon of use of animals as mere things (in contemporary factory farming, for example), nor are they more accepting of such objectification. (Barry Lopez is another poignant example, as we have seen.) The crucial difference is that they distinguish between "respectful" and merely exploitative relations to animals on the basis of a more primal relatedness of all animal life. "Respect" for prey animals does not require that one leave them alone, but, among other things, that one hunt them respectfully (an oxymoron for animal rights thought), where "respect" derives its meaning from awareness of a mutual participation in the ebb and flow of life itself, and from awareness of both the dignity of the animal within the process and the dignity of the process itself.

> *The killing and eating of animals by hunting-gathering peoples is not seen as victory over a reluctant nature, not as an assertion of will or virility, but instead as part of the larger gift of life, a receiving from the hand of a conscious power according to the state of grace of the recipients. The crucial moment in the hunt is not the "taking" of a life but the moment of respect and affirmation for a giving world* (*Ibid*, 45–46).[9]

The distance from Taylor could not be greater. For Taylor, bowing to the necessities of life requires the use of priority principles to free ourselves from the *prima facie* duties based on his concept of "respect" for nature, and Taylor's moral agent would have to will to be freed from such conflicts. The ideal of harmony is the ideal of avoiding such conflicts. Shepard's hunter affirms the very intertwining of life in death, death in life, as the essence of a giving, *living* world. From this perspective, it is precisely detachment that is the danger.

> *We are easily seduced by our own empathy because of our fear and outrage at the indifferent destructiveness around us. However, kindness towards animals demands a true sense of kinship. To be kindred does not mean we should treat animals as our babies. It means, instead, a sense of many connections and transformations—us into them, them into us, and them into each other from the beginning of time. To be kindred means to share consciously in the stream of life that goes beyond the solace of a grasped emotion. It takes you out of, not into, yourself* (*Ibid*, 46).

From Shepard's perspective, Taylor's account of appreciation of organisms as teleological centers of life suffers from two fatal flaws. To the extent that it is modeled on scientific knowledge, it is the attitude of a detached observer, and as such the emotion it tends to evoke easily slips into sentimentality—the illusion of participation (Taylor's "sense of oneness with all other living things") by one who is essentially (and ideally) a nonparticipant. Second, in its abstract focus on individuals it marginalizes the system of interdependence that is the very key to the gift of life. Knowingly or unknowingly, the result is an affirmation of "life" in abstraction from the reality of life. In this sense, Taylor is closer to Schweitzer's abstract mysticism than it seems at first glance, in spite of the obvious differences between them.

One result of these two flaws is that any appropriation of natural life by a human being is easily viewed as an implicit assertion of human superiority. Just as the traditional claim that human beings metaphysically transcend nature involves a hierarchical view of the world, in which human beings are the pinnacle and meaning of creation, so the disengaged "biocentric" view takes human beings out of nature, now as a moral obligation. Any involvement with nature that is not sanctioned by the priority principles—which are not to presuppose human superiority—is a reimposition of the old hierarchical view.

TAYLOR'S EXISTENTIAL APPEALS—THE EXIGENCY OF DISENGAGEMENT

Taylor's own existential appeals, which appear at each stage of his unfolding of the meaning of "membership," underline these points. Although Taylor's ideally competent evaluator is objective and disinterested, he or she is not, it seems, without empathy or feeling. Indeed, it is precisely these capacities to which Taylor appeals. Thus (a) when we become aware of our own dependence on biological and physical requirements, we realize that we are *members* of the Earth's *Community* of Life in the sense of being biologically dependent on it.[10] But (b) we can also realize that "we are in the *same existential situation* as are all other living things" (Taylor 1986, 105; my emphasis), since they too have a good of their own. The objective and disinterested stance yields knowledge about ourselves: we *share* the same existential situation with all other living things, and for Taylor this commonality gives existential import to the notion of shared membership in the earth's community. This is not just an objective fact about us, but concerns the very context in which we undertake to live our lives, including our moral lives. In light of the biocentric outlook, I choose as a member of a community. This is intensified when (c) we view ourselves as sharing a value, the value of freedom in the sense of lack of constraint. A new level of community is constituted by this sharing (cf. *Ibid*, 111). The fourth step (d) returns the focus to us in our relatively unique status as newcomers on the stage of evolution. We are newcomers to the community of shared values, and there is here an implied appeal to a sense of decorum. The historical background against which this is directed is the theological view that human beings have been created in the image of God, and thus (uniquely) have inherent value. Everything else would then have merely instrumental value for humans. The carefully built up existential sense of community is inconsistent with this view of mastery. Finally, (e) when we realize that while we are dependent on the community, it is not dependent on us, and that our disappearance might even be beneficial to many species, we get a dramatic restatement of (a).

From the points of view of both Schweitzer and Shepard, it is striking that these existential claims are the result of objective knowledge, not of reflection (however saturated by knowledge) on our engagement in the process of life. This carries over to Taylor's discussion of the second component of the biocentric outlook, the natural world as a system of interdependence. Here Taylor emphasizes that life is "a tightly woven web" (*Ibid*, 116),[11] in which human interference tends to be destabilizing and destructive. Given the senses of membership already developed, when we "see ourselves as an integral part of the system of nature" (*Ibid*, 117), "integral" has to have the normative import of being nondestructive and nondestabilizing.[12]

This tendency toward a *prima facie* existential disengagement is again strengthened in Taylor's discussion of the third component of the biocentric outlook, viewing organisms as teleological centers of life. This is an elaboration of the recognition that all organisms have "a good of their own." Here the point is developed in terms of the ability of scientific inquiry, which is generally aimed at establishing general laws, to give us insight into unique individual organisms. I suggest above that there are other paths to such insight, but for Taylor the fact that scientists do not, as such, view animals in terms of some function they might play in our lives would be an important virtue. This capacity to take up the standpoint of another living thing expands our specifically *moral* imagination: "Seeing them as we see ourselves, we are ready to place the same value on their existence as we do on our own. If we regard ourselves as possessing inherent worth, so will we be disposed to regard them" (*Ibid*, 128). Once again, the move is from a detached insight to an existential attitude. But the existential attitude that is used as a point of departure is itself detached: our regard of ourselves as possessing inherent worth. Our lived perspective is played down in favor of a detached recognition of (our own) inherent worth. Further, the existential attitude we arrive at is not only detached, it is also abstract. The individuals whose inherent worth we accept are viewed in abstraction from their concrete context, which is the web of interdependence.

Finally, Taylor again couples existential disengagement with existential appeal in the fourth component of the biocentric outlook, the denial of human superiority. The background is again the traditional, often theologically based, anthropocentrism of the Western religious and philosophical traditions. If we were to reject those traditions, if we were to reject superiority in *those* senses, "A total reordering of our moral universe would take place" (*Ibid*, 134). Since the tradition involves recognition of the inherent worth only of human beings, abandoning that tradition requires that we recognize equal inherent worth: the "principle of species-impartiality" (*Ibid*, 155).

> *This is the principle that every species counts as having the same value in the sense that, regardless of what species a living thing belongs to, it is deemed to be prima facie deserving of equal concern and consideration on the part of moral agents. Its good is judged to be worthy of being preserved and protected as an end in itself and for the sake of the entity whose good it is* (*Ibid*, 155).

The detachment of this "judgment" becomes clear when we ask just who it is that is judging in this manner. The answer is that it is any individual to the extent that that individual takes up the perspective of a detached observer, leaving aside everything else. The individual making the judgment disappears precisely in his or her individuality and biological reality. One might well find this kind of detachment out of place in a biocentric outlook. From *this* perspective, overthrowing traditional anthropocentrism can only mean giving consideration to the good of every other organism equal (in the sense of the same) to the consideration I give to my own. This can only mean that I must *prima facie* not harm any other living being, and this is Taylor's conclusion. Of course I, as a concrete existing individual, cannot live without violating this *prima facie* duty, so priority principles are needed.

TAYLOR'S RULE OF FIDELITY AND HUNTING

In this light, Taylor's rules of nonmaleficence, noninterference, and fidelity are a false extension of human ethics, an expression of sentimentality. The rule of fidelity is of particular interest here, since Taylor's most detailed discussion of hunting, trapping, and fishing is found in his section on the rule of fidelity. Taylor's rule of fidelity involves duties

1. Not to break a trust that a wild animal places in us (as shown by its behavior);
2. Not to deceive or mislead any animal capable of being deceived or misled;

3. To uphold an animal's expectations, which it has formed on the basis on one's past actions with it;
4. To be true to one's intentions as made known to an animal when it has come to rely on one (*Ibid*, 179).

It is clear that if one accepts these duties, hunting and fishing are *prima facie* forbidden. In discussing Taylor's arguments against hunting, philosopher Theodore Vitali writes, "I take this argument to be patently absurd because no such trust relationship exists in any way, shape, or form in nature. If ever there were an example of imposing a human value upon nature, Taylor takes the cake here. George Schaller has clearly shown that deception and ambush is an evolved practice of hunting for virtually all stalking animals, of which the human animal is an example" (Vitali 1993, 74, ftn. 9).[13]

From the perspective of Taylor's concept of "respect," Vitali's objection is simply beside the point. Ethical principles and rules cannot be simply "read off" of nature. Nature is not a model for the proper actions of a moral agent. Here again, the tension in Taylor's thought begins to make itself felt. On the one hand, human beings belong to the world of culture, and "the natural world" is defined in terms of the absence of human intrusion or control (cf. Taylor 1986, 3). Here "nature" is opposed to "culture." Yet it is an essential component of the biocentric outlook *on nature* that "we see human life as *an integral part of the natural order of the Earth's biosphere*" (*Ibid*, 101, Taylor's emphasis). Where do we look for guidance when we want to define what "respect for nature" might mean: to nature or to culture? to the nature that is defined by the absence of human intrusion and control or to that "natural order" of which human life is an integral part? or to the cultural order itself? Vitali's charge is that Taylor is reaching into the cultural order, into the realm of human ethics, to get his sense of "fidelity," and that it is natural for him to do so only because he has excluded human beings from "the natural world" to begin with. Given that exclusion, no sense of the appropriateness of human action vis-à-vis the natural world can be oriented in terms of that world itself, since human interaction with "the natural world" is by definition "intrusion or control," even when it is justified by the priority principles.

But there is an obvious alternative, one Taylor skirts by leaving the system of interaction out of his development of his concept of "respect." In "the natural order" as a system of interdependence, plants and animals are always vulnerable to being eaten. The natural order cannot function, has no order, without these relations of appropriation. Why should a respect for nature that draws its nourishment from viewing ourselves as an integral part of the natural order, that natural order that is a web of interdependence, of the eater and the eaten, propose a rule of fidelity that is so patently at home in the world of person-to-person actions?

If, instead of defining "the natural world" in opposition to human "intrusion," Taylor had set out to deal with the moral relations that hold between humans, creatures of culture, and the *wild*, then he would have had to deal with the fact that human beings have been an integral part of the natural world in North America, the world of the wild as opposed to domestic animals and plants, for over ten thousand years. This is not just a fact, but also an important, and appropriate, aspect of human reality.[14] Put in Taylor's terms, if human beings are an integral part of the natural order, and if the natural order is a web of interdependence, then it makes little sense to develop a moral principle that requires human beings to remove themselves from the natural order to the greatest extent possible. Rather, the problem is to determine *appropriate* ways of being part of the natural order. This appropriateness has to have one foot in culture, since only a self-reflective, cultural being can even ask what is and is not appropriate, and one foot in the natural, as opposed to the cultural, world, since we are a part, an integral part, of that world, and the question is how we can live, how we can be interdependent, in an appropriate manner.

Wild animals do not have a natural trust in human beings. The fact that some wild animals do not fear human beings is not the same thing as a trust that the animal places in us. To go to the Galapagos and have the experience of being surrounded by animals exhibiting no fear of human beings is to have the experience of being, for the most part, totally and completely irrelevant to those animals. To walk up to a booby sitting on its nest is to experience being ignored in a very extreme form. Indeed, one is not being ignored, one's presence simply does not count, and as the bird swivels its head one has the feeling that it is looking right through one. One is not really present until one gets very close (two feet or less) from the nest, at which point the bird may jab at the intruder with its beak. Historically, humans have taken advantage of this lack of fear and have slaughtered such birds. Today most of us feel a general revulsion at such practices, and would generally feel that they are not consistent with an attitude of respect for nature. (A hunter-gatherer would not necessarily agree.) But note that this is completely irrelevant to issues concerning hunting.

All hunting has to do with game animals that have a natural wariness, speed, or other characteristics that make them appropriate as game. They are animals that have evolved under the pressure of predation, both animal and human. And when a wild animal has become acclimated to human presence and expects, for example, to be fed, no ethical hunter will shoot it, and for a hunter this would be an expression of respect for the animal's (compromised) wildness. Laws against baiting fields fit in here. Unlike Taylor's rule of fidelity, such respect does not forbid bugling an elk in rut or calling in a wild turkey gobbler during the spring mating season, but it does put constraints on what counts as hunting. There is an ethics of hunting, one that is firmly rooted in respect for the animals hunted (cf. Posewitz).

It is, however, a legacy of the inherent value/instrumental value and anthropocentrism/biocentrism distinctions as drawn by philosophers such as Singer, Regan, and Taylor, namely as mutually exclusive dichotomies, that discussions of hunters' ethics can so easily fall into the pattern of looking at an individual rule or principle, and asking whether it is anthropocentric or an expression of biocentrism (of respect for nonhuman species or individuals).

Thus, in Brian Luke's discussion of the rules of hunters' ethics he writes that "it might seem that [the principle of fair chase] embodies some respect for the hunted animals, that the hunters are trying to be fair *to their targets* by refraining from totally overwhelming them with technology" (Luke, 27).[15] But Luke argues that this can be at most a "superficial fairness, insofar as it is not in the interests of any particular game animal to be hunted 'sportingly,' but rather not to be hunted at all" (*Ibid*, 27). Luke quotes Joy Williams reducing the idea of "fairness" in fair chase to the idea that hunting is "a balanced jolly game of mutual satisfaction between the hunter and the hunted—*Bam, bam, bam, I get to shoot you and you get to be dead*" (Williams, 252; quoted at Luke, 27).[16] The either/or of anthropocentric instrumentalism versus genuine respect for wild animals is very clear. Given this model, the only way to interpret the notion of "fair" in "fair chase" is as an anthropocentrism disguised by anthropomorphism—hunting as a game in which hunter and hunted participate out of mutual consent and for mutual benefit. Only on that assumption can the concept of "fair play" as it applies to human interaction be applied to hunting.[17] The result is that notions like "fair chase" cannot be expressions of any genuine respect for animals (Luke, 28). Given his presuppositions, Luke's conclusion follows.

As Theodore Vitali has recognized, the alternative is to recognize that in analyzing the concept "fair chase," the notion of fairness as it applies to contests between human beings, in which "each participant has a rightful claim to be treated fairly by the other participant," a right that is based on the right to autonomy, is simply inapplicable (Vitali 1996, Part I, 21). Understood in this manner, fairness has its proper place in the ethics of human-to-human relationships. Williams is right to make fun of the idea that hunting is a game freely entered into by both hunter and hunted, but it is a mistake to think that her sarcasm hits a serious target. Vitali argues that "fair chase" has nothing to do with a "fair chase contest" (*Ibid*). Its field is rather that of "the causal relationship of hunting pleasures to human virtues and the latter in turn to human excellence" (*Ibid*, 24). Fair chase is a matter of the "virtues of self-imposed discipline and restraint," the meaning of which is to be found in

> *a very unique kind of wisdom, an enduring insight into the most fundamental of all relationships, the relationship of kinship of all living beings on this planet . . . the ultimate community of nature. As a result of this insight, this root or radical wisdom, the hunter gains a profound attitude of respect for and a sense of gratitude to, even humility before, the animals he or she hunts and the natural order that nurtures and sustains all, including him or herself* (*Ibid*).

One may reject this "wisdom," but to do so calls for something other than the rhetorical instruments of sarcasm. We are at one of those points at which it is quite legitimate for anyone to reply, "I don't get it." But as Schweitzer realized, this is no argument; indeed, this is not a field for argument, in the strict sense. This does not mean that we cannot discuss these issues, discuss them deeply, and that discussion may not have a deep effect on our attitudes and ways of acting in the world. But this is not a field for universal principles and their simple application to cases, to anyone, anywhere, anytime.

AN ETHIC OF PURITY?

I have suggested that our obligations to nature are always correlates of our sense of self. Thus, my critique of Taylor's account of our duties to nature needs to be complemented by a look at Taylor's understanding of the human self as it relates to the natural world. I have been arguing that Taylor's concept of "respect" is inconsistent with the biocentric outlook on life as he lays it out, that it rests on an abstraction from the fact of interdependence. The question that now has to be raised is whether there is a similar abstractness at work in his idea of what it means to be a concrete human being living in and from the world. If one assumes that any harm to a nonhuman organism by a moral agent, if that harm is not justified by something like Taylor's priority principles, is an implicit assertion of human superiority, this will be in effect to accept Taylor's concept of "respect." What is the implied concept of the human self, such that it should conceive of itself as fundamentally foreign to the system of interdependence that is constitutive of the natural order?

While it is clear that Taylor is not adopting an attitude of negation of life, I think that it is interesting to compare his position to the *ahimsa* commandment—the commandment not to kill and not to harm living creatures—as it is found in the Hindu traditions of Samkhya, Yoga, and Jainism. According to Albert Schweitzer's interpretation, Jainism represents a step forward from both Brahmanism and the Samkhya doctrine, which emphasized "being exalted over the world" (Schweitzer 1952a, 79), since Jainism places great importance on ethics. But when Schweitzer investigates the origin of the *ahimsa* commandment, he argues that it does not develop out of compassion [*Mitleid*] because it ignores "the giving of real help" (*Ibid*, 80). "The commandment not to kill and not to harm does not arise, then, from a feeling of compassion, but from the idea of keeping pure [*Reinbleiben*] from the world" (*Ibid*, 80). In other words, the *ahimsa* commandment is based on an ideal of purity, rather than an ideal of being morally engaged in and with the world. It is a development of the negation of world and life characteristic of Indian thought in general (*Ibid*, 80–81).

More recent interpretations of the *ahimsa* commandment, as it appears in the Samkhya and Yoga traditions, emphasize that the issue is less an ethical than a spiritual one, with the ethic being subservient to the spiritual goal of liberation of the spiritual self from nature

(Jacobson, 299; cf. also Doniger). This spiritual attitude was a drastic revision of the outlook of the *Brahmana* texts (ca. 800 B.C.E.), in which the universe is seen as a hierarchy of foodstuffs, with the more powerful eating and dominating the less powerful (Jacobson, 290–293). The *ahimsa* commandment is the result of replacing that hierarchy with a different hierarchy, that of the pure-impure. The goal of purity is the goal of renouncing one's place in the hierarchy of eater-eaten (*Ibid*, 293). "Human beings by identifying themselves with the food chain misunderstand therefore their true identities" (*Ibid*, 297). Jacobson contrasts this attitude with both the later ethical teaching of duties to the whole as found in the *Bhagavadgita* and contemporary environmental concerns. In the *Bhagavadgita*, "humans have the duty to participate in this activity [of eating as participation in the wheel of interdependencies of beings] for the sake literally of holding the world together (*lokasamgraha*)" (*Ibid*, 299). Dan Gerber quotes *The Upanishads*:

> *Oh wonderful, oh wonderful, oh wonderful,*
> *I am food, I am food, I am food,*
> *I am an eater of food. I am an eater of food. I am an eater of food*
> (quoted by Gerber, 44).

Similarly, contemporary environmental concern, as exemplified by Aldo Leopold, "is based on an understanding of the interdependency of humans with the community of all living beings and a wish for an integration of humans with nature" (Jacobson, 300).

Now it is clear that Taylor's attitude of respect for nature is far from the negation of life found in Samkhya and Jainism since Taylor's respect for nature leads to recognition of the *prima facie* duty to promote or preserve any being that has inherent worth. Taylor does not restrict himself to not killing or harming. His principle requires that one engage in the world, promoting and preserving the well-being of living things. And yet, Taylor's position is more similar to Jainism than one would expect. Rather than requiring that we intervene in natural processes in order to promote and protect the well-being of living organisms, Taylor's rule of noninterference requires "a general 'hands off' policy with regard to whole ecosystems and biotic communities, as well as to individual organisms" (Taylor 1986, 173). This rule is based primarily on the concept of individual organisms as teleological centers of life and the idea of freedom that

Taylor developed in analyzing the concept of being a member of the Earth's Community of Life (*Ibid,* 173–174, 105–111). However, the following should be noted. To say that an organism is a teleological center of life is to say that it has a good of its own, and this means that we can evaluate events in the world as being good, bad, or indifferent to the well-being of that organism. But I have argued above that to say that something bad happens to an organism (a duck is killed by a fox) is not to make a moral statement. All we can say is that something (nonmorally) bad has happened to the duck, and something (nonmorally) good has happened to the fox. The question is why this would be any different when a human being kills a deer and eats it—on the assumption that the human being was not cruel, acted in ways that respected the wildness of the animal, and ate the meat of the animal that was killed.

There are two correlative reasons at work here. The first is found in Taylor's examples of naturally functioning ecosystems and of the ways in which human beings disrupt, damage, and destroy them (*Ibid,* 116–117). The strong suggestion is that as systems of interdependence, natural systems function *properly* in the absence of human influence. It is of course true for Taylor that when we see the natural world as a system of interdependence, "we will see ourselves as an integral part of the system of nature" (*Ibid,* 117). But given the human potential for disruption, damage, and destruction of natural systems, along with the attitude of respect for the organisms that live in these systems, the rule of noninterference naturally enough commands that we stay out of naturally functioning systems as much as possible. Human presence and influence threatens their proper functioning and contaminates their inherent worth, their purity. The correlative motive is found in the purity of the attitude of detachment analyzed in the last subsection. If what is most essential to our humanity is our capacity to be objective and disinterested evaluators, if this is what constitutes us as moral agents, then the attitude of theoretical detachment that yields such objectivity and disinterest will have a practical counterpart: respect for nature as the requirement that moral agents remain pure from the world as a system of interdependence. In other words, an ideal of purity need not be rooted in negation of life. It can also be rooted in a form of "respect for life," but only when the life of the moral agent is viewed in abstraction from the fact that all life, including human life, the life of human beings as moral agents, is intrinsically a system of interdependence. Given such an assumption, any way of life that is one of living interaction with and assimilation of the life around us must be held to be *prima facie* immoral. The morality of respect, as Taylor understands it, is a morality of purity.

This analysis can be pursued one step further. Taylor assumes that, for a human being as a moral agent, to kill an animal when this is not sanctioned by his priority principles is an expression of the presumption of human superiority. The problem is thus one of the imposition of a hierarchy of values. As I have suggested above, however, relations of eater-eaten are not necessarily hierarchical relations in which the higher or more valuable properly consumes the lower or less valuable. Such assumptions of hierarchy were at work in the *Brahmana* texts and their system of sacrifices (Jacobson, 289–291). But in addition to the individualistic focus of the *ahimsa* commandment in Samkhya and Yoga, these traditions also contain a nonhierarchical view of ecological interdependency. A striking example is found in Vacaspati Misra's *Tattvavaisaradi* (900 C.E.), a commentary on the *Yogasutra*, which contains the following passage:

> *For the human body is sustained by the use of bodies of tame animals, of birds, of wild animals and of plants. Similarly bodies like the tiger (are sustained) by the use of the human bodies and those of tame and wild animals and of others. And again in the same way the bodies of tame animals, the birds, and wild animals (are sustained) by the use of plants and similar things. Likewise the divine body (is sustained) by the use of sacrifices, of goats and deer and the flesh of other animals, of ghee, and baked rice cakes and of branches of mango and handfuls of darbha grass offered by human beings. In the same way the deity also sustains human and other beings by granting boons and showers. Thus the dependence is reciprocal* (*Tattvavaisaradi*, 177; quoted by Jacobson, 300).

What is striking about this text is that in contrast to the emphasis on hierarchy and dominance in the *Brahmana* texts (Jacobson, 290), the emphasis is on reciprocity, on reciprocal dependence.[18] Humans belong not just to the order of eaters, but also naturally enough to the order of the eaten. There is nothing unnatural about a tiger eating a human being, no violation of a natural hierarchy. The goal continues to be that of escaping the entire system of interdependency. If, however, one rejects the fundamentally negative attitude toward life and the world, this nonhierarchical view opens up the possibility of asking what a proper and respectful way of living in the community of living things, of being an integral part of that community, might be.

TAYLOR'S RESPONSE: HUMAN BEINGS AS "PART OF NATURE" AND HUMAN BEINGS AS MORAL AGENTS

The critique of Taylor's concept of respect for nature presented above hinges on the claim that when he develops the concept of the moral attitude of "respect for nature" on the basis of acceptance of the biocentric outlook on nature, Taylor ignores the most crucial part of the biocentric outlook, namely recognition of the fact that all living things, including human beings, are part of an interdependent web of life. I argue that giving interdependence a central role in determining the proper sense of "respect" is necessary if the position is to be coherent. At the very least, "respect for nature" must include respect for the web of interdependence and for the place of human beings as part of this web.

After Paul Taylor graciously agreed to read a draft of Part III, he wrote an extensive reply in which he very respectfully discusses, and rejects, my critique. His response centers on what he calls the "threefold nature of human beings": humans are *animals*, *valuers*, and *moral agents*. "As animals with physical bodies, we are situated in and cannot escape from the complex web of interdependence that sets the context for any living thing's survival and well-being" (Taylor 2000a, 1). As valuers, "the values we choose express our 'sense of self' insofar as they represent our deepest motivations and ideals. They constitute our answer to the question, 'What is the meaning of life?' (*Ibid*). As moral agents, "we are also beings who consider some of the things we do as being right or wrong. We regard ourselves as not being at liberty to do as we please, to pursue our values, when the lives of others are made better or worse by what we do" (*Ibid*, 1–2).

Against the background of these distinctions, Taylor argues that we have to ask again what it means to say that humans are part of nature. As animals, we are of course part of nature. "But we are not 'part of nature,' we do not 'participate' in nature, insofar as [we] are valuing beings who pursue values of our own choosing. Nor are we 'part of nature' when, as moral agents and judges, we place ourselves under the authority of principles we autonomously adopt as our own moral norms" (*Ibid*, 2) In short, *causally* we are part of the web of interdependence, but *conceptually*, qua valuers and moral agents we are autonomous, and are thus *as such* not 'part of nature.'

Taylor emphasizes this distinction between the *causal* and the *conceptual* in his discussion of intrinsic value in his "Response."

> *The kind of beings that can have intrinsic value occur in their existence are living things, species of living things, communities of living things, and systems of living things (ecosystems). These are all entities (I shall call them "biological entities") that have a good of their own. They are such that it is possible to benefit or harm them, independently of benefiting or harming something else. This is a* conceptual *point, not a* causal *claim.* Causally, *we cannot actually benefit one organism without our causing death to another, except in cases of symbiosis . . . That any such entity does carry on its existence in a condition of realized well-being is a state of affairs that has intrinsic value. It is a good thing that such an entity is in such a condition of existence, independently of whether any other biological entity is in such a condition* (*Ibid*, 7, 8).

The question, "Why should anyone believe that the realization and continuation of a biological entity's state of well-being is intrinsically good?" goes "to the very roots of environmental ethics" (*Ibid*, 8).

Finally, in the "Response" Taylor argues that contemporary sport hunting is different from the hunting of hunter-gatherer cultures, and that this is a crucial difference.

> *The hunter in modern life (as distinct from the Stone Age hunter-gatherer)* pursues a human value when she chooses to be a hunter. *When she values hunting because of what hunting means to her, even when she takes its meaning to be that she is "participating" in the system of interdependence (life-requiring-death), hunting in that way for that reason is a matter of the hunter's free, autonomous choice . . . It is because modern (sport) hunting is a matter of free choice that it can be judged morally . . . Given the foregoing considerations, one cannot argue that hunting is morally permissible because it is a way of participating in nature, or because it is a way of*

understanding ourselves to be a part of the system of interdependence (Taylor, *Ibid*, 2–3).

Taylor is arguing here against a conflation of human beings as *valuers* and human beings as *moral agents*. Even if a human being values hunting as a way of participating in nature, even if a human being finds that certain forms of hunting add richness, meaning, and depth to his or her life, this begs the question of the moral permissibility of hunting. Giving a positive, existential meaning to the hunt does not answer the moral question.

I am Kantian enough in my own thinking about human ethics (cf. Part I) to agree that these are deep and important distinctions. I disagree with Paul Taylor concerning the way he uses the distinctions. As the quotes above show, he applies the distinction between the conceptual and the causal in two different, though related, contexts: the context of the moral agent and the context of the intrinsic value of biological entities. I argue that analysis of the two contexts yields results that converge.

The first context concerns the status of the human being as moral agent. In this chapter, I have argued against Taylor's understanding of the attitude of respect for nature on the grounds that he uses this concept to take human beings as moral agents out of nature, at least as a moral ideal. Taylor's position is not a direct result of the distinction between the human being as animal, as valuer, and as moral agent. His position is rather a moral teleology that Taylor argues is the rational result of the moral agent's moral reflection. By the same token, I do not deny the nature/culture or the animal/moral agent distinctions, but it is important to recall that a moral agent is also part of the natural world. The very fact that we are moral agents is a product of evolution. The only way to avoid this would be to embrace either a Cartesian dualism or a Kantian metaphysics of phenomena versus noumena. Dualism is an expression of traditional anthropocentric metaphysics and, contra Kant, I argue elsewhere that the moral point of view does not necessitate a strong distinction between phenomena and noumena.[19]

The central question in environmental ethics is how we *as moral agents* are to live *as part of the natural world*. Our being as animals and our being as moral agents can and must be carefully distinguished, but they cannot be simply separated from one another. When we, qua moral agents, begin to think about the ethical principles that are to shape our relations to the natural world, the fact that we are a *part of* that world is crucial to our deliberation from the very beginning, not something that can be brought in at a later stage as an empirical fact that merely determines the conditions under which the principles we have already arrived at must be applied. The very fact that we *are* essentially part of the web of interdependence makes a specifically *environmental* ethic both necessary and possible in the first place. (Oddly enough, this means that Taylor's ethic is not really an *environmental* ethic at all. It is not an ethic of our relationships to the world that surrounds *us*.) Moral thinking cannot begin by considering our relationship to the natural world from a purely detached, disinterested, extra-worldly attitude: here the moral subject, purely as such, there the teleological centers of life (we ourselves among them) that are to be treated with respect; here the respecter, there the respectees. A veil of ignorance about our place in the web of interdependence is fundamentally misguided here. That is why I find Taylor's biocentric outlook on nature, which *begins* with the belief that humans are members of the earth's community of life and the belief that humans are part of the web of interdependence, such a powerful perspective from which to think about our fundamental attitude toward nature, wild nature. But precisely something like such a veil of ignorance descends on the biocentric outlook on nature when Taylor begins to develop his attitude of respect toward teleological centers of life. In contrast, I argue that one must begin one's reflections on environmental ethics by recognizing that human beings as moral agents are part of the web of life. This is not to naturalize environmental ethics, it is to ecologize and existentialize it.

The second context in which Taylor misapplies the distinction between the causal and the conceptual concerns the good of individual living things, species, systems, and so on. Taylor's use of the distinction between the *conceptual* possibility of considering the benefit and harm to a given biological entity independently of the harm or benefit to other biological entities, and the *causal* necessity that such benefit generally involves correlative harm to other biological entities, is indeed important. As Taylor argues, the conceptual possibility is fundamental to recognition of the intrinsic value of biological entities, and this means that the conceptual possibility is the key to establishing the moral considerability of biological entities. The good of any teleological center of value can be, and must be, given moral consideration. Giving such consideration is an essential aspect of what it means to respect biological entities. But to establish moral considerability is not yet to establish a general moral principle of a concrete moral attitude. In Kenneth Goodpaster's terms, to establish moral considerability is not yet to establish moral *significance*.[20]

I argue above that the very concept of a teleological center of life *presupposes* the causal dimension that Taylor wishes to exclude as being conceptually irrelevant to the construction of his basic moral principle. To stop with a general concept of moral considerability is to stop with an abstract concept of a teleological center of life, and this yields an abstract concept of respect. The fact that Taylor must immediately bring in priority rules to make concrete decisions and moral action possible is evidence of this. But once the web of interdependence is acknowledged as fundamental to the very concept of a teleological center of life, and thus as fundamental to the moral significance of biological entities, as it is when the biocentric outlook on nature as a whole is taken as the point of departure, the conceptual point as Taylor presents it, although it is of crucial importance, can be seen for what it is, an abstraction. It is an important abstraction, but as such it is an inadequate perspective from which to develop a moral theory, especially an environmental ethic. A substantive concept of "respect" cannot be based on the conceptual abstraction.

If one thinks about environmental ethics from the conceptually distinctive perspective of a human being as a moral agent, but without acknowledging *and affirming* (or respecting) the causally constitutive fact that moral agents are part of the web of interdependence, then an environmental ethic committed to the ideal of removing moral agents from nature, from the web of interdependence, is a foregone conclusion. Similarly, if we think about environmental ethics from the point of view of the good of a particular biological entity (which for that reason has intrinsic value) independent from the fact that the good of that particular entity is causally bound up with the goods of other biological entities, all that we can do is bring in this causal connection as an unfortunate constraint on our ability to realize the moral principles that were based on the abstraction. The web of interdependence is relegated to a strictly secondary status. I argue that recognition and affirmation of the fact that we, along with every biological entity, are part of the natural world of interdependence has to be part and parcel of—the very framework of—our cultivation of an appropriate attitude of respect for the natural world. It is precisely because we are part of the natural world that we, *as* moral agents, deliberate about our moral place in nature. The resulting concept of "respect" must also involve respect for both nonhuman biological entities as parts of the web of interdependence and for *ourselves* as part of the web of interdependence.

I argue that this objection holds against Taylor's theory as it is developed in *Respect for Nature*, with its strictly individualist understanding of intrinsic value, and against Taylor's comments in his "Response." But even if this claim is rejected, I still argue that when Taylor acknowledges, as he now does, not only the intrinsic value of individual living things, but also the intrinsic value of species, communities, systems, and so on (cf. Taylor, 1994), he does not just *expand* the scope of his environmental ethics as he understands it; he fundamentally *changes* the ethical principle that can be based on this broadened range of intrinsic value. This is true *even if one sticks to Taylor's insistence on the priority of the abstract good of particular biological entities*. The *conceptual* point about the good of an individual living thing is immediately confronted by the *conceptual* point about the interrelated good of that very thing's species, the interrelated good of the community of which it is a part, the interrelated good of the particular ecosystem within which it exists, and so on. The interrelatedness of these various goods is not just a contingent empirical truth, but is also a matter of the structure of the realm of intrinsic value. This means that it is not axiologically coherent to isolate conceptually the good of the individual living thing from the good of the species or system. I argue above that the causal truth is built into the very concept of a "teleological center of life." My point here is that the conceptual point cannot be understood independently of the causal truth because the causal truth is built into the relevant concepts of "community" and "system" as loci of intrinsic value.

It is worth noting that Kant's concept of respect for the dignity of persons allows him to consider a moral agent as "a member in a possible realm of ends, to which his own nature already destined him" (Kant 1785, 52). Taylor's concept of respect for the intrinsic value of a teleological center of life considered independently of the web of interdependence does not allow us to consider an analogous realm of natural inherent value of which a moral agent would be a member. Taylor's analogue to Kant's realm of ends is his ideal of harmony, but again, the thrust of that ideal is to remove human beings from nature, not to integrate human beings into the web of life in a way that affirms human membership. In these terms, the task in environmental ethics is to think of the web of interdependence as a starting point for constructing an analogue to the realm of ends for persons. The web of interdependence is not such an analogue simply by virtue of its biological reality, even though human beings qua animals are part of the web. It becomes thinkable as an analogue only from the perspective of an appropriate attitude of respect. Such a position is clearly closer to Aldo Leopold's conception of human beings as citizens of the land-community (Leopold, 204) than to Taylor's moral ideal of the removal of human beings from the natural world.

Obviously one can consider the good of or benefit to a given living thing without considering the harm to other living things that is causally connected to this good or benefit. The crucial question concerns the role this fact should play in environmental ethics. Taylor attempts to base his *environmental* ethic on this possibility. The lesson to be learned from the biocentric outlook on life is that such an environmental ethic is environmentally or ecologically naïve, and this means that it is not genuinely biocentric. An environmental ethic that is cut off from the very structure of the natural world not only fails to be persuasive, it is not a truly environmental ethic. I suggest taking the dilemma of individualism versus holism by the horns. My position avoids the Scylla of pure individualism, which makes a genuinely environmental ethic impossible, and the Charybdis of a pure holism, which leads to the charge of "environmental fascism" because it ignores the intrinsic value of individuals. Taylor is on the right track when he recognizes the intrinsic value of *all* biological entities, but the result of this recognition is a dramatic transformation of his ethics of respect for nature.

When the distinction between human beings as animals and as moral agents, and the distinction between consideration of the inherent value of a particular biological entity and consideration of that entity as part of the causal web of interdependence, have been reinterpreted along the lines I suggest, Taylor's argument against "modern (sport) hunting" has to be reexamined. As I interpret his point in the "Response," he argues that even if one grants that there is a genuine human value in the experience, such value has no justificatory value at all if nonsubsistence hunting is immoral. But if my critique of Taylor's use of the two crucial distinctions is correct, then questions of meaning become central. If the task of environmental ethics is to formulate principles in terms of which human beings actively and affirmatively participate in the web of interdependence, then the meaning of various human practices becomes crucial. Hunting is an activity that can be pursued in many different ways and with many different attitudes toward both the practice and the prey. Some practices are clearly the expression of an anthropocentric attitude of domination. Others involve respectful participation in the web of life even as life is taken and given.

I take up some of the issues involved here in Part IV.

CONCLUSION

If one rejects Taylor's kind of detachment, along with the resulting disengagement and tacit commitment to an ideal of purity, one must take up the problem of engaging with and appropriating the living things of the natural world, and of doing so with *respect*. To insist that life is appropriation, that human life is inevitably bound up with the death of living things, is not necessarily to return to traditional anthropocentrism. The alternative to the ideal of purity is to develop a concept of respect that *integrates* the four elements of Taylor's biocentric outlook. I discuss this in more detail in Part IV.

In our culture we need to think about the notion of "respectful participation" in the natural world of which we are a part, even in this postindustrial, postmodern, increasingly virtual life-world in which we live. Paradoxical as it might sound, the problem with Taylor's ethics of respect for nature is not that it contains a residual anthropocentrism, as critics such as Sterba have charged, but that it is not genuinely *biocentric*. Interdependence is a fundamental fact of life because life as we know it is essentially consumptive or appropriative. All life, of whatever form, exists and continues to exist only because it appropriates energy in one form or another from its environment. *Life is appropriation.* Any ethical theory that does not recognize *and affirm* this fundamental fact is not a serious candidate for an environmental ethic, no matter how many principles it contains for *overriding* our supposed *prima facie* obligation *not* to appropriate.[21]

Our goal in developing an environmental ethic should not be to place limits that are as restrictive as possible on human interactions with nature and on human appropriations of nature. This purely quantitative approach is a result of taking human beings as moral agents out of the web of interdependence. The goal is rather to recognize these interactions as encounters with an other to which we are essentially kin and to insist on the dignity of both ourselves and the other in these relations, even and especially when the relations are appropriative. In short, the goal is to give a properly respectful form to our relations with nature, wild nature. Substantive restrictions will then emerge naturally from such "respect."

Part IV

Respect for Nature and Biocentric Anthropocentrism

Chapter 6

Biocentric Anthropocentrism

INTRODUCTION

In Parts I–III I have discussed, in greater or lesser detail, four positions that arrive at some strikingly similar conclusions, even though their philosophical points of departure are deeply divergent. Two of these, those of Tom Regan and Peter Singer, are explicitly programs of "ethics by extension" in the sense Holmes Rolston III introduced into the contemporary debate: they take a moral status that is commonly granted to human beings and argue that it must be extended to a fairly broad range of animals. A third position, that of Paul Taylor, may be extensionist in the sense that environmental ethics is modeled on human ethics, but Taylor avoids extensionist arguments. It may be that only Albert Schweitzer develops an approach that is not extensionist in any sense. Again, the fundamental philosophical principles invoked differ dramatically. Singer is a utilitarian. Regan is a rights theorist. While Taylor has reasons for not using the word "rights" beyond the human realm, his theory is in many ways a broadening of Regan's approach. Schweitzer's mysticism of the infinite will to live and the principle of reverence for life which he thinks is a "necessity for thought" move in a different philosophical world from any of the other three, though it is tempting to view Taylor's principle of respect for nature as a (fairly radical) revision of Schweitzer's principle.

In each case, my discussion of the position has been accompanied by a critique. The power of these critiques should be weighed carefully. As I have already noted, in Part I my critique of ethics by extension is limited, first to sketching an alternative approach to human ethics that cannot be meaningfully extended to cover animals, and then to opening up an alternative way of thinking about the human relationship to wild animals. It goes without saying that this discussion will not convince anyone committed to Regan's or Singer's position. The door to an ethics by extension program remains open if one simply rejects the sketched (Kantian) approach to human ethics. What Part I does do, I think, is to take away the sense of inevitability that their arguments may (and are intended to) evoke. Once an alternative becomes visible, it is possible to explore it and develop arguments for it, which is what I do in Parts II and III.

In the case of Schweitzer (Part II) my critique is, I think, internal and thus more fundamental and more telling. The basic charge is that Schweitzer fails to live up to two of his own demands, namely the demand that any ethic that is to be a candidate for adoption must be life affirming and the demand that the mysticism that is a "necessity for thought" avoid the abstract mysticism of the tradition. Schweitzer's ethic of reverence for life is, of course, by definition life affirming, but a closer look reveals that the concept of "life" on which his ethic is based is a one-sided abstraction. Schweitzer's concept of "life" does not refer to life as we know it, to life that by its very nature is intertwined with death and in which our own mortality is part of what makes us alive. Schweitzer can only affirm life as opposed to death, with the result that his ethic cannot affirm the concrete phenomenon of life, as we know and live it, in which life and death are intertwined. I argue that his position thus falls back into the abstract mysticism he attempted to overcome.

In the case of Taylor (Part III) my critique is that, while Taylor claims that his principle of respect for nature is supported and made intelligible by what he calls the biocentric outlook on nature, it turns out that the principle is supported and made intelligible by at most three of the four components of the biocentric outlook. If one takes the pivotal second component seriously—the belief that all living things are integral elements in a system of interdependence (Taylor 1986, 99–100)—then the biocentric outlook no longer supports Taylor's specific concept of "respect for nature." I have further argued that one cannot avoid this conclusion, and that if this is the case, dramatic revisions of Taylor's concept of "respect" are necessary. Here too the critique is internal and thus, if successful, more fundamental and telling.

Against the background of these critiques, each section also contains a discussion of an attitude of respect that I argue is of great importance to the way we experience and think about the relationship between human beings and the broader world of which we are a part. In Part I, I develop a notion of "respect" for the animal world, using

Barry Lopez's claim that "The aspiration of aboriginal people throughout the world has been to achieve a congruent relationship with the land, to fit well in it" (Lopez 1987, 297) as a point of departure. Without engaging in a nostalgic romanticism about native peoples, I believe that this ideal of a congruent relationship remains important in the contemporary world.[1] Indeed, Aldo Leopold's concept of the "citizen of the land community" is in many ways a secularized, demythologized version Lopez's concept of "transcendent congruency" (*Ibid*).[2] The problem it poses for us is that of developing cultural practices that help us achieve such congruent relationships, and of developing a culture that is imbued with such practices. From this perspective, if one hunts with the proper attitude and understanding of the meaning of one's practice, hunting can be one way of achieving this kind of congruence.

Lopez's model of a respectful relationship between humans and wild animals is both interactive and appropriative. When human lives are lived in true intimacy with wild animals, even appropriative relations are more than simply events of killing and eating; rather, they are permeated by a sense of meaning and decorum, of rightness and wrongness in human relationships with what is other. The act of hunting, killing, and eating a wild animal can be an expression of human participation in and affirmation of the process of life itself, the process that gives life, the process in which life is intertwined with death. I argue that the goal is not to distance ourselves as much as possible from this process, but to learn to participate in it in meaningful and affirmative ways, ways that respect the integrity of both human and nonhuman individuals, and of the encompassing systems of which they are a part. The test of the practices in which this goal is to be realized will have to be complex. As a start, it would have to involve the integrity of the forms of life we live, and of the world we live in and from, the way in which our sense of ourselves is integrated with the various kinds of otherness that surround us and in one way or another permeate our lives, and the affirmative meaning or "respect" that is given to those kinds of otherness in our relations with them.

Whatever the truth of his claims, Lopez is very clear that the aboriginal model, with its "contracts" and resulting sense of respectful participation in the natural world, based as it is on the practices of hunting and gathering peoples, and on the myths that pervade their lives and give meaning to their actions, is no longer appropriate. We live in a world permeated by scientific and technological objectification, and by the economic commmodification of animal life and the natural world in general. Lopez therefore calls for a new contract, not a return to the old one. He leaves us with the question of what form such a new contract might take.[3]

In Part II, I argue that Albert Schweitzer's principle of reverence for life, once it is freed from the one-sided concept of "life" that is characteristic of his mysticism, can provide a foundation for an affirmative view of the concrete process of life, in which life lives from life, and for giving affirmative meaning to human *participation in* this process as an expression of reverence both for the process and for the individual lives that are lived within the process. Against Schweitzer's abstract mysticism of the infinite will to live, I argue that once we adopt an affirmative attitude toward the concrete process of life as a whole, the appropriation of individual life need not be an expression of disrespect or lack of reverence, but rather can be a positive expression of respect or reverence for both the process and the individual life appropriated.

Once Schweitzer's principle of reverence for life is made concrete in this manner, it allows us to begin to differentiate a series of questions that often get confused with one another in debates about the morality of hunting. As a beginning, I distinguished the following questions:

1. Is nonsubsistence hunting ethically permissible in principle?
2. Does *my* appropriation of the principle of reverence for life allow hunting?
3. How can one kill such beautiful creatures?
4. Why do we hunt?

In addition, Schweitzer's insistence on individual responsibility in making practical decisions in concrete situations, but always guided by the principle of reverence for life, has the virtue of making those who submit to the discipline of this principle increasingly thoughtful and mindful of life in all forms in their actions and practices. From this perspective, *why* one hunts and *how* one hunts are crucial to the moral dimension of contemporary hunting. Schweitzer's refusal to offer a system of rules for governing such practical decisions, something like Paul Taylor's priority rules, is essential to the cultivation of such mindfulness. Schweitzer insists that recognition of the subjectivity of such decisions is necessary if we are to take genuine responsibility for our actions. Yet this is not the subjectivity of "anything goes."

In Part III, I argue that Paul Taylor's concept of "respect for nature" is not adequately grounded in his own account of the "biocentric attitude on nature." Once human membership in the earth's community of life as a system of interdependence is taken seriously *and affirmatively*, the path is open for developing a concept of "respect for nature" centered on the respectful *participation* of human beings in natural processes. My crucial point is that Taylor is wrong in claiming that the justification of human appropriation of nature requires overriding a *prima facie* obligation not to harm. I argue that the problem environmental ethics has to face is that of determining what a *respectful* appropriation of nature would be.

The result in each section is a rejection of positions that require that human beings take themselves out of natural processes to the greatest extent possible. My objection is not that this ideal cannot be realized or that not enough allowance is made for practical realities, but that it is a false ideal. I argue that the position that begins to emerge from these considerations is truly *biocentric* and *egalitarian*.[4] Indeed, it is more genuinely biocentric than the positions called "biocentric egalitarianism" in recent environmental ethics literature, in part because it is definitively freed both from the necessity of primarily defining itself in terms of a contrast to traditional anthropocentrism[5] and from the constant temptation to import principles from human ethics into the sphere of environmental ethics.

ANTHROPOCENTRISM REVISITED: THE CASE FOR BIOCENTRIC ANTHROPOCENTRISM

I have just noted that many contemporary forms of "biocentric egalitarianism" have defined themselves to some extent negatively, that is, by the need to overcome traditional anthropocentrism. I suggest that this is what has made their claim that human beings must be removed from nature seem so plausible to many. I have therefore begun to articulate an alternative that affirms the human place in nature, but without falling back into traditional anthropocentrism.

There is, however, an obvious objection to my approach. The objection is that while my position rejects the claim that the function of nonhuman nature is to satisfy human needs and desires, by affirming human appropriation of nature, especially wild nature, I remain committed to a kind of anthropocentrism—practical anthropocentrism, perhaps, as opposed to metaphysical anthropocentrism. The objection is that when I argue that a human can take the life of a wild animal and do so with "respect" for both the individual animal and for the system of which the human and the animal are parts, I have justified the human domination of nature. The claims about "respect" would then be mere window dressing. In other words, as Paul Taylor argues, the fact that all species of animal life appropriate energy and nutrition from their environment does not justify that human beings, who are moral agents, take the life of sentient creatures or destroy living nature when this is not truly necessary. If this objection is valid, we are faced with an unfortunate pair of options: EITHER we embrace an anthropocentrism in which human beings view themselves as being entitled to use other species in any way they see fit,[6] OR we embrace a biocentrism that removes humans from nature to the greatest extent possible.

One way of putting the objection would be to say that when I attempt to take seriously the belief that nature is a system of interdependence, and to *affirm* the fact that all life, including our own, is appropriation, I am basing environmental ethics on biology or ecology. Such a move to base value on fact, ethics on science, is generally acknowledged to be invalid, and if that is what I am doing, then my project is a failure and what I have emphasized as existential affirmation merely serves to disguise the fallacy. I will now investigate this issue by taking a critical look at botanist William H. Murdy's biology-based proposal of a "modern version" of anthropocentrism. A consideration of Murdy's position is of value in this context because it allows one to see precisely where the lines are drawn between scientific considerations and existential analysis.

Murdy argues that biological reality requires that we accept anthropocentrism not only as a biological fact of life, but as a fundamental aspect of the way we choose to orient ourselves in the world. This would seem both to violate the fact-value distinction and to fall back into the fallacies of traditional anthropocentrism, in spite of the shift of emphasis from religion or metaphysics to science. But Murdy claims to be offering a third alternative, not just a modern version of traditional anthropocentrism. My thesis is that Murdy's "modern anthropocentrism" is in close agreement with the position I have been developing, for a careful reading shows that his position does not in fact violate the fact-value distinction and that his anthropocentrism is more genuinely biocentric than the positions I criticize above.

Murdy begins his essay "Anthropocentrism: A Modern Version" by contrasting traditional anthropocentrism, "the idea that nature was created to benefit man" (Murdy, 316; Murdy quotes Xenophon's *Memorobilia*), with the implications of the Darwinian revolution. "Modern" thus means being scientific in orientation. From this perspective, Murdy writes, "Species exist as ends in themselves. They do not exist for the exclusive benefit of any other species. The purpose of a species, in biological terms, is to survive to reproduce" (*Ibid*, 316).[7] For Murdy, this biological view requires that we reject traditional, theologically-based anthropocentrism, which is to be replaced by a new, biologically-based anthropocentrism.

> *To be anthropocentric is to affirm that mankind is to be valued more highly than other living things in nature—by man. By the same logic, spiders are to be valued more highly than other things in nature—by spiders. It is proper for men to be anthropocentric and for spiders to be arachnocentric. This goes for all other living species* (*Ibid*, 316–317).

From this perspective, a so-called biocentric egalitarian approach that demands that we human beings give equal consideration to the interests of each living organism of whatever species makes little *biological* sense.

> *I may affirm that every species has intrinsic value, but I will behave as though I value my own survival and that of my species more highly than the survival of other animals or plants. I may assert that a lettuce plant has intrinsic value, yet I will eat it before it has reproduced itself because I value my own nutritional well-being more highly than the survival of the lettuce plant* (*Ibid*, 318).

In other words, we may claim to take the intrinsic value of nonhuman nature seriously, but reality teaches us that this is strictly window dressing with no practical value in the long run. We may give lip service to such a principle, but ultimately we cannot actually live our lives in conformity with it.

This can be interpreted as an application of the "ought implies can" principle that one cannot have a moral duty to do something that one cannot do. If a truly egalitarian approach runs contrary to our nature as biological organisms, Murdy seems to be saying, then it cannot be a valid ethical principle. From this perspective, the very phrase "biocentric egalitarianism" would be a contradiction in terms, since it is based on an idealistic misinterpretation of "*bios*," of life itself.

Murdy does not reject the concept of intrinsic value as it applies to nonhuman nature. "An anthropocentric attitude toward nature does not require that man be the source of all value, nor does it exclude a belief that things of nature have intrinsic value" (*Ibid*, 318). This follows from the same insight that gives rise to Murdy's anthropocentrism, namely insight into the biological "purpose" of any species, but it still means that in the final analysis it is *our* intrinsic value that is crucial *to us*, simply because it is *ours*. But Murdy also thinks that we need to *cultivate* recognition of such intrinsic value—for strictly anthropocentric reasons. Thus, "man should acknowledge the intrinsic value of things; otherwise, he will not have sufficient motivation for ecological survival" (*Ibid*, 318). We thus have the seemingly paradoxical recommendation that we recognize the intrinsic value of nature and act in light of that recognition for strictly instrumental reasons of enlightened self-interest—an anthropocentrically based biocentrism.

How far has Murdy come from traditional anthropocentrism? How new is his "modern version"? Murdy argues that the philosophical and theological traditions of anthropocentrism are based on the first, but only the first, of two crucial steps. Quoting David Bohm, Murdy writes, "'Man's first realization that he was not identical with nature' was a crucial step in evolution, writes Bohm, 'because it made possible a kind of autonomy in his thinking, which allowed him to go beyond the immediately given limits of nature, first in his imagination, and ultimately in his practical work'" (*Ibid*, 318; quoting Bohm, 18). As I point out in the Preface, the notion that human rationality makes human beings superior to and the purpose of the rest of the world, goes back at least to the ancient Greeks, and determines much of the Western philosophical tradition. Kant's traditionally anthropocentric interpretation of this "first realization" in his "Speculative Beginning of Human History" is so pregnant that it is worth quoting in full.

> *[Man] discovered in himself an ability to choose his own way of life and thus not to be bound like other animals to only a single one. The momentary delight that this just discovered*

> *advantage may have awakened in him must have been followed immediately by anxiety and unease as to how he should proceed with this newly discovered ability, for he knew nothing about its hidden characteristics and distant consequences. He stood as if at the edge of an abyss; . . . and it was now equally impossible for him to turn back from his once tasted state of freedom to his former servitude (to the rule of the instincts)* (Kant 1786, 51).[8]

For Kant, it is this step out of the world determined by instinct that gives human beings lordship over the rest of nature.

> *[Man understood] himself (though only darkly) to be the true end of nature, and in this regard nothing living on earth can compete with him. The first time he said to the sheep, "the pelt that you bear was given to you by nature not for yourself, but for me;" the first time he took that pelt off the sheep and put it on himself (Gen. 3:21); at that time he saw within himself a privilege by virtue of which his nature surpassed that of animals, which he now no longer regarded as his fellows in creation, but as subject to his will as means and tools for achieving his own chosen objectives (Ibid, 52–53).*[9]

But for Murdy, this nonidentity with nature, which motivates Kant's traditional anthropocentrism, is only the first step, and unless it is followed by a second step, it leads to destruction. Murdy quotes botanist Hugh H. Iltis: "Not until man accepts his dependency on nature and puts himself in place as part of it, not until then does man put man first. This is the greatest paradox of human ecology" (Iltis, 820; quoted by Murdy, 318).[10]

To say that recognition of human dependence on nature is the rationale for recognizing the intrinsic value of nonhuman nature sounds suspiciously like a "noble lie" in the tradition of Plato: we should teach our children and our culture (and ourselves) to believe that nature has intrinsic value, but the real reason why this belief should be taught and made a principle of action has to do with the well-being of human beings. Nature is important in the final analysis, so it would seem, because it is crucial for human survival and flourishing, not because it has intrinsic value; the acknowledgement of intrinsic value would be important for its instrumental value. Murdy's balancing act is precarious, and it seems ultimately to fail. In an attempt to avoid sacrificing the intrinsic value of nature on the altar of traditional anthropocentrism in his "modern" anthropocentrism, Murdy makes the point that recognition of the intrinsic value (or inherent worth) of non-human nature does not contradict a properly understood anthropocentrism. But the fact remains that as long as the point of departure for thinking about the human relationship with nature is found only in biology, nature's value for human beings is in the final analysis instrumental. After all, *we* are the ones thinking about the importance of nonhuman nature. We are the ones who have to decide how we will act. For Murdy's biologically-based anthropocentrism, this means that *we* are the ones who really count—for us. But that means that in human terms, *we* are the ones who really count, period.

Yet, there is something more than biology at work in Murdy's thought when he goes on to say, "Personal identification with greater wholes is essential to the discovery of our own wholeness" (Murdy, 322). This is crucial, and easy to miss. This is no longer merely a biologically-oriented point about the inescapably instrumental value of the natural world to us (as to any species). It is rather a point about what our status as part of the natural world means for our sense of selfhood. Murdy's concern shifts from the level of biological and evolutionary fact, which is his home territory as a botanist, to that of *meaning* and the structure of self, which is his home territory as a human being. Scientific theory gives way to existential reflection and analysis. This meaning cannot be simply read off of biological fact, as Schweitzer saw clearly, though this is what Murdy at times seems to want to do. It is an issue not for detached, objective thought, but rather for the self-reflection of that being that, within the constraints of its biological nature, has to choose how it is to live and what the meaning of that life is to be.

As I state at the beginning of this book, our attitudes toward and our obligations to nature are always the correlates of our sense of self. Murdy's strength is his insistence that our sense of self—which includes our sense of ourselves as moral agents—not be cut off from the fact that we are part of the biological world. But when this fact is integrated into our sense of self as reflective moral agents, it makes possible and necessary a sense of identification that is based on a passionate participation. Taylor loses this by replacing it with a detached sympathy. Odd as it may sound, properly understood, only Murdy's kind of "modern" anthropocentrism is truly biocentric. Rather than the paradox of an anthropocentrically-based biocentrism, Murdy's attempt to integrate Darwinian insight with existential insight yields a biocentrically-based anthropocentrism.[11]

In such a biocentrically-based anthropocentrism, in which human beings are recognized to be part of nature both biologically and existentially, rather than being transcendent to nature, recognition of the intrinsic value of nature is not a mere window dressing or a useful illusion. There is indeed a practical point here: if one insists on uncoupling recognition of the intrinsic value of nonhuman nature from the self-serving structure of organisms and species, especially *Homo sapiens*, one dooms one's ethic to being ineffectual for reasons found in Taylor's own biocentric outlook on nature. But the deeper point, as I argued in Part III, is that intrinsic value cannot be separated from instrumental value. When Murdy writes that recognition of intrinsic value does not change my anthropocentric behavior, this is license to domination only when read in terms of traditional anthropocentrism. Read in terms of Murdy's "modern," genuinely *bio*centric anthropocentrism, it leads us to recognize and affirm ourselves as a part of the world we live in and from, and calls us to be a responsible and respectful part of that world—for the sake of the world and of ourselves. Self and world cannot be thought independently of one another. Self-respect and respect for nature therefore become correlates of one another. The self is situated in such a way that the world it is part of is an issue to that self in its very sense of its selfhood: Murdy's appeal to biology raises existential questions. Biology or ecology cannot answer these questions; only we can.

NATURE AND CULTURE

The concept of "respect" that these considerations point to also allows us to cut the Gordian knot of some recent debates that turn on the distinction between the "natural" and the "cultural." Thus, Holmes Rolston III argues,

> *The killing and the eating of animals, when they occur in culture, are still events in nature; they are ecological events, no matter how superimposed by culture. Humans are claiming no superiority or privilege exotic to nature. Analogous to predation, human consumption of animals is to be judged by the*

> *principles of environmental ethics, not those of interhuman ethics* (Rolston 1988, 79).

Rolston argues that since the killing and eating of animals are most fundamentally events in nature, the principle that should apply is the principle of the nonaddition of suffering (Ibid, 59–60). Human killing of animals should not "amplify the cruelty in nature" (Ibid, 59).[12]

I argue above that hunting need not presuppose human superiority to nature (though some hunters' behavior does express such an attitude). This is not, however, because human hunting is an "ecological event" *as opposed to* a cultural event, but rather because *in addition to* being an "ecological event" (which it is) it can be a cultural expression of respect for nature.[13] Indeed, Rolston's way of approaching killing and eating, as well as hunting, leaves him vulnerable to valid objections. In a response both to Rolston and to Ned Hettinger's friendly amendments to Rolston, Paul Veatch Moriarty and Mark Woods argue that Rolston's equation of natural predation and human predation (hunting) cannot be sustained "because hunting and meat eating by humans do not have the same status as natural events that animal predation has as an integrated part of a natural ecosystem. We argue that meat eating and hunting as they occur in our society are not properly considered as natural events" (Moriarty and Woods, 395–396). If this point is granted, they argue, then Rolston's principles are not sufficient. We have to examine cultural practices themselves, as cultural practices:

> *If the goal in question is one of participating in the natural processes of nourishment practiced by our gatherer-hunter ancestors, then it seems that this value could be achieved with minimal harm by berry picking. If, however, Hettinger and Rolston choose to define the goal as one of participating in the natural process of predation, then we must ask why one should adopt this goal rather than a similar goal (i.e., participating in the natural process of food gathering) which can be achieved with less harm* (*Ibid,* 397).

Moriarty and Woods thus insist "there is nothing natural about meat eating and hunting in our culture. . . . As cultural activities, they involve a different set of moral duties" (*Ibid*, 399).[14]

Moriarty and Woods ask first if it still makes sense to speak of our participating in natural processes. "But do any of us living in mainstream Western society really participate in the 'logic and biology' of any natural ecosystem?" (*Ibid*, 410; quoting Rolston 1988, 60—not Hettinger, as they state). Beyond that challenge, they "believe there is a *prima facie* case against hunting and meat eating" in any culture (Moriarty and Woods, 402), although their focus is on hunting in Western culture. In their brief look at hunting by European-Americans over the past two centuries, they identify four categories: subsistence hunters, market hunters, sportsmen, and landowners for whom wild animals are a nuisance. They view sport hunting as a form of recreation in which wild animals are "recreational resources" (*Ibid*, 403). They thus conclude, "Not only will Western culture continue to thrive in the U.S. with or without hunting, but the cultural history of hunting provides no moral justification for the continued practice of hunting today" (*Ibid*, 403).

This approach is inadequate on a number of levels. Just as there are levels of naturalness (cf. *Ibid*, 399, ftn. 22), so there are levels of participation in the logic and biology of natural ecosystems. Moriarty and Woods's rhetorical question, "But do any of us . . . really participate . . . ?" has to be confronted with the reality of such lives as those of Ted Kerasote, Richard Nelson, Mary Zeiss Stange, George Bird Evans, and countless others, who do in various ways and to differing degrees participate intensely in the functioning of their ecosystems. In addition, as I note above, to define sport hunting as recreation or as killing for pleasure begs the question. If Moriarty and Woods had started by defining sport hunting as hunting for its own sake (cf. Luke 1997, 25, ftn. 2), they would have had to investigate the different possibilities covered by the "for its own sake." Participation is one of these, and the fact that our lives are generally far from such participation is a compelling reason to change the way we live, not an argument against the cultural justification of hunting. My approach to the issue from the point of view of respect for the nature that we are still part of, rather than from the point of view of an animal rights ethic that tends to remove us from nature, casts a very different light on the questions Moriarty and Woods raise. They are correct to insist that nonsubsistence human hunting must be justified as a *cultural* practice, but they fail to show that this cannot be done because they systematically denigrate the value of affirmative participation in natural processes.

Hunters need not denigrate the value of berry picking (or photography, as Ortega y Gasset does; cf. also A. Jones, 24–27, 39–41; I discuss this issue in more detail below)—in fact many of the most passionate hunters are berry pickers, mushroom gatherers, bird watchers, wildlife photographers, etc.—but once the value of participating in natural processes of appropriation is acknowledged, the decision to pick berries *rather than* to hunt is a decision of personal authenticity (which has a place for personal inclination), to be made from the attitude of respect for nature or reverence for life. Not everyone needs to hunt, but this is not a matter for universal moral legislation. Berry picking should not be used to maintain the fiction of a life lived outside of the web of interdependence, of life and death. For example, in Montana the picking of wild huckleberries harms bears that depend on the berries for food. Whether one picks them oneself, or buys them in a store (it is a big business), one is not free from the impact caused by the practice.[15]

RESPECT FOR NATURE AND ECOFEMINIST VEGETARIANISM

No discussion of the (or an) attitude of respect for nature today should avoid the challenges that come from some forms of ecofeminism. Not every form of ecofeminism has been opposed to all forms of hunting (Karen Warren and Mary Zeiss Stange have already been mentioned above), but many ecofeminists have taken vegetarianism to be an important component of any defensible ecofeminism. In particular, in *Neither Man Nor Beast: Feminism and the Defense of Animals*, Carol J. Adams has argued that "an ideology that ontologizes animals as usable" is "the result of a human-animal dualism that is embedded within a racist patriarchy" (Adams 1994, 15).

The link between the domination of women by men and the domination of animals by humans, both justified by the ideology of patriarchal culture, is deeply embedded in the Western philosophical tradition. Adams quotes Wendy Brown in the epigraph to her Preface:

> *It was precisely the sharpness of the Athenian conception of manhood that bore with it a necessary degradation and oppression of women, a denial of the status of "human" to women. To the extent that women were viewed as part of the*

> *human species, they would recall to men the species' animal or "natural" aspect. Alternatively, women could be denied fully human status and remain the somewhat less threatening repository of the "lower elements" of existence. Seen in this context, Aristotle's infamous characterization of women as "deformed males" bears significance as more than incidental misogyny. Aristotle does not merely posit the general inferiority of women but describes them as "incomplete beings," their thinking as "inconclusive," and the female state in general as a condition of "deformity and weakness." Women are also depicted as "matter" in need of the "form" only men can supply. Women are therefore not merely lesser humans than men but less-than-human, malformed, and ill-equipped for the human project, creatures in a gray area between beast and man* (*Ibid,* 10; quoting Brown, 56).

For this (our) tradition, the dualisms man/woman and human/animal are of the same mold (cf. also Bergman).

In her earlier book *The Sexual Politics of Meat*, in which she analyzes the manner in which our culture's language constructs animals as food, Adams develops the concept of "the absent referent" (Adams 1990, 40f.). Specifically in English, the animal and its meat are often referred to by two different terms—cow/beef, pig/pork, deer/venison, sheep/mutton—with the result that not only does the once living animal whose corpse is now being eaten become absent, the fact that one is eating a dead animal is made absent. "The absent referent permits us to forget about the animal as an independent entity" (Adams 1994, 17). Even when this is not the case, as with chicken, lamb, and others, our language resorts to mass terms such as "meat." "Objects referred to by mass terms have no individuality, no uniqueness, no specificity, no particularity" (*Ibid*, 27), thus again allowing us to avoid the fact that we are eating something that was a living individual. Our very language protects us from the reality of our practices.

Given the general relationship she sees between the domination of women and the exploitation of animals, it is hardly surprising that Adams denies that hunting can be reconciled with ecofeminist ethics. But she is aware that not all ecofeminists agree, some out of a refusal to absolutize, and she gives a respectful hearing to claims that some forms of hunting seem to escape the taint of the absent referent.

> *Killing animals in a respectful act of appreciation of their sacrifice, this argument proposes, does not create animals as instruments. Instead, it is argued, this method of killing animals is characterized by relationship and reflects reciprocity between humans and the hunted animals. Essentially, there are no absent referents. I will call this interpretation of the killing of animals the "relational hunt"* (Ibid, 102–103).

Since my own position clearly falls under Adams's concept, a discussion of her critique of the relational hunt is important.

Adams's critique of the relational hunt has two stages. First, she notes that the relational hunt requires "ontologizing animals as edible" (*Ibid*, 103). This turns animals into instruments: "animals' lives are thus subordinated to the human's desire to eat them even though there is, in general, no need to be eating animals" (*Ibid*).

The second stage of Adams's critique involves applying what she calls "the method of contextualization." This requires that "the context describing how we relate to animals" be provided, and, writing in the early 1990s, Adams finds that this has not been done for the relational hunt. In other words, as I understand Adams, there is no systematic account or analysis of the context of the relational hunt beyond the anecdotal level. Adams argues that when the context is examined, inconsistencies begin to appear. Thus, while proponents of the relational hunt claim for it a relationship of reciprocity, Adams replies, "reciprocity involves a mutual or cooperative interchange of favors or privileges. What does the animal who dies receive in this exchange?" (*Ibid*, 104). Adams considers the experience of sacrifice as one possibility here, but dismisses it as unverifiable. But once the willingness of the prey is put in question, a connection between the relational hunt and what Adams calls the "aggressive hunt" arises—"the eliding of responsibility or agency" (*Ibid*).

Using Ortega y Gasset as a representative of the aggressive hunt, Adams detects an evasion of responsibility for killing in Ortega's claim that "To the sportsman the death of the game is not what interests him; that is not his purpose. What interests him is everything that he had to do to achieve that death—that is, the hunt" (Ortega y Gasset, 105). Adams responds, "The erasure of the subject in this passage is fascinating. In the end the hunter is not really responsible for willing the animal's death . . . In the construction of the aggressive hunt, we are told that the killing takes place not because the hunter wills it but because the hunt itself requires it" (Adams 1994, 104). This is parallel to the conceit that in the relational hunt the animal gives itself. In both cases, responsibility for the taking of life is avoided.

Adams makes two final points. First, the understanding of the relational hunt is often based on the proponents' understanding of Native American hunting practices, and Adams charges that this amounts to "cannibalizing" the practices of another culture (*Ibid*, 105). Finally, Adams notes that the relational hunt cannot be practiced on a large scale in the contemporary world. She concludes, "The problem of the relational hunt is that it is a highly sentimentalized individual solution to a corporate problem: what are we to do about the eating of animals? We either see animals as edible bodies or we do not. The hunting issue therefore is ultimately a debate about method" (*Ibid*, 106).

My response to Adams's critique of the relational hunt is brief. A full response would require an extended discussion of both *The Sexual Politics of Meat* and *Neither Man Nor Beast*, and this cannot be done here. But the more immediate limitation, as Adams says, is that at the time of her writing there was little in the way of a developed theory of the relational hunt, and she does not refer to any discussions of this practice. Karen J. Warren's brief discussion of Sioux hunting practices at the end of her essay, "The Power and Promise of Ecological Feminism" (Warren, 145–146), may be in the background. Today, one would have to make reference to Mary Zeiss Stange's *Woman the Hunter*[16] and Val Plumwood's important essay "Integrating Ethical Frameworks for Animals, Human and Nature: A Critical Feminist Eco-socialist Analysis," as well as to such texts as Ted Kerasote's *Bloodties*, Richard Nelson's *Heart and Blood*, Jim Posewitz's *Beyond Fair Chase*, and Allen Jones's *A Quiet Place of Violence*, which are not ecofeminist. Stange, Kerasote, and Jones make serious attempts to explicate the context of what Adams calls the "relational hunt." It would be speculative to pose Adams's views against these later theories, so I make only four comments on Adams's critique of the "relational hunt."

First, I have argued above that the concept of "respect for nature" as I have developed it does not presuppose a hierarchy in which humans are viewed as being superior to animals. "Ontologizing animals as edible" does make animals instrumental in human nutrition and cultural life, but I show above that instrumental use and recognition of inherent value are not mutually exclusive. Many writers on the "relational hunt," whether they draw on Native American traditions (Warren, Nelson) or not (Kerasote, Stange, Jones), deny the charge of domination or subordination. There is, of course, the sense in which any use of a living organism subordinates the other's interests to the interests of the user. But, as Adams insists, the important question concerns the *ideology* of such subordination (Adams 1994, 15).[17] It is striking that when one reads the work of Ted Kerasote, Richard Nelson, Mary Zeiss Stange, Allen Jones, George Bird Evans, and many others, one finds anything but the "erasure of the referent." What one finds instead is the description and analysis of a set of practices that cultivate a recognition or a reverence for a specificity of time, place, and event in which the grace and beauty of the prey animal is present at every point in the process from hunting, through killing, and finally the eating of the animal.[18] It is no accident that Ted Kerasote makes it a practice to speak of, e.g., "the elk *whom* I've been eating since October" (cf. Kerasote 1994, 182; my emphasis) and "I put *him* [the elk Kerasote had killed] in my mouth and began to feel the land pass through my body" (*Ibid*, 176, my emphasis; cf. also 221), and to develop practices that individuate the elk he hunts and kills (*Ibid*, 245–246). Val Plumwood goes beyond any of these attitudes and practices when she holds that respectful hunting requires that the hunter be in the role not only of eater but also of potential eaten (Plumwood 2000, 314; cf. A. Jones, 27–28).

While I do not find myself in agreement with a number of aspects of Ortega's philosophy of the hunt, the passage Adams quotes does not represent the erasure of the subject. It is rather the outline of an analysis of a *practice*, one that situates the meaning of the kill within that practice. One may reject Ortega's analysis of the practice, or one may reject the practice Ortega analyzes, but his analysis should be recognized for what it is. The problem is not that Ortega describes and analyzes a practice without attention to its context, but rather that the practices he analyzes are often practices that have lost much of their context. His account of "blood lust" and descriptions of the automatic urge to pursue and kill (Ortega, 91–92, 118–119), are descriptions of practices that have lost contact with their place in the organic web of life. The real problem has to do not so much with Ortega's description and analysis of these practices, but with the specific practices themselves. Paul Shepard's praise of Ortega to the contrary, the practices Ortega describes fall rather under Kellert's category of the "sport/dominionistic" hunter than the nature hunter.

Adams is correct in taking Ortega's philosophy of hunting as a philosophy of the "aggressive hunt" as opposed to the "relational hunt."

Second, Adams argues that there is no genuine "reciprocity" in the relational hunt. But she is looking for reciprocity in the wrong place. While the notion (or experience) that a prey animal "gives" or "offers" itself to the respectful hunter is part of some traditions, it is not universal (cf. Kerasote 1994, 225–226). Many accounts of the relational hunt emphasize reciprocity not on the one-to-one level of hunter (and eater) to prey animal, but on the level of the overarching process of life of which each is a member. Thus, the Sioux grandparents' instructions to their grandson, in the account Karen J. Warren cites, include the instructions:

> *Offer also a prayer of thanks to your four-legged kin [deer] for offering his body to you just now, when you need food to eat and clothing to wear. And promise the four-legged that you will put yourself back into the earth when you die, to become nourishment for the earth, and for the sister flowers, and for the brother deer. It is appropriate that you should offer this blessing for the four-legged and, in due time, reciprocate in turn with your body in this way, as the four-legged gives life to you for your survival* (quoted by Warren, 146).

Salient here are both the idea that the prey "offers" its body (although this "offering" is not overtly psychologized—the deer is not presented as a willing sacrifice) and the idea that reciprocity takes the form of a return of the human body to the generative cycle of life and death. This is also the dominant motif in Part III of Kerasote's *Bloodties*, which is titled "Webs." Kerasote requires a "daily intimacy with country" as a condition for respectful hunting, "the condition that I've come to believe is necessary if one is to receive life instead of merely take it" (Kerasote 1994, 192).

"Receiving" in this sense is not so much a matter of the individual prey animal deliberately giving itself of what Paul Shepard calls "the moment of respect and affirmation for a giving world" (Shepard 1991, 46). In an essay that reached me as I was completing this manuscript, Val Plumwood answers Carol Adams's question as to what the animal who dies receives from the "exchange": "The answer is that it has already received it in life itself, existence as part of the cycle of embodiment exchange. The idea of the food chain as a cycle of sharing and exchange of life in which all ultimately participate as food for others is what we should understand by reciprocity here" (Plumwood 2000, 319, n. 21).

Third, Adams suggests that appeals to Native American traditions in support of the idea of the "relational hunt" amount to "cannibalizing what is presumed to be their hunting model" (Adams 1994, 105).[19] Certainly there are serious issues concerning what can amount to a colonial appropriation and occupation of carefully selected aspects of a romanticized Native American spirituality. But not all proponents of the relational hunt make appeals to Native American tradition a central part of their theory, and some who have used certain Native American traditions as models have done so after serious study and what can only be called apprenticeship. One cannot reasonably charge Richard Nelson (or Gary Snyder) with cultural cannibalism (cf. Nelson 1983, 1994, 1997). In addition, Adams has no objection to appealing to the practices and traditions of other societies, as long as the appeal supports her own position. Thus, she writes, "Why not hold up as a counterexample to ecocidal culture gatherer societies that demonstrate humans can live well without depending on animals' bodies as food?" (Adams 1994, 105). This presupposes one has already accepted Adams' position, and that all alternatives to that position are "ecocidal." The dangers of cultural imperialism and romanticism need to be taken seriously by all sides.[20]

Fourth, Adams correctly notes that the relational hunt cannot be practiced today on a large scale. Few of us can (and even fewer do) live the way Kerasote, Nelson, and Stange live.[21] As I noted above, some advocates of the relational hunt take the position that their only personal alternative to the relational hunt would be vegetarianism (e.g., McGuane). Adams is correct in stating that the relational hunt cannot in any direct sense solve the moral problems raised by the meat industry. But as the examples of McGuane, Kerasote, Buddy Levy, and others show, it is often reflective hunters who actively oppose the meat industry and the alienation from the natural world that is part and parcel of the "corporate problem." It is also clear that there is a lot of room in the United States today to practice more hunting with the attitude of respect and reciprocity of the relational hunt than is currently the case. By the same token, few of us can grow all the food we need for a vegetarian diet (and even if we could, the habitat loss would have be taken into account). Ted Kerasote makes this point in recounting his experience the first time he killed an elk.

> *I bore her heavy quarters down through the forest, one by one, her gift at last accepted, her pain not pushed away. After all, where could her pain go? To the next county? To the next state? Or perhaps to the arctic where the oil needed to transport rice and beans to Wyoming, equivalent nutritionally to the meat of these yearly elk, spills and ends the lives of three otters, a half-dozen seals, and a score of common murre chicks, which is how I reckoned the costs of being a fossil fuel vegetarian* (Kerasote 1994, 185).

The point here is not that it is our oil-based economy, or more broadly Adams's "corporate problem," that justifies Kerosote's kind of hunting. The point is that there is no human life that can be lived without the responsibility for the burden it imposes on the lives of the animals around us, and that respectful hunting is one way of living with and living up to this responsibility, of giving it an affirmative meaning in the practices that constitute our lives. One can, of course, follow Schweitzer and claim that all killing of animals, even when it is necessary for bare survival, is evil. This would mean that hunting should never be given a positive meaning within a culture, even when it is necessary for survival, since that would suppress the guilt that individuals and groups must take upon themselves in order to live. I argue above that this kind of position is based on an abstract conception of life and leads naturally to an abstract mysticism of (abstract) life.

If one takes as point of departure a robust concept of life, and affirms both the goodness of the system of life of which every living thing is part and product, and the potential for goodness in our own lives within this process, this changes the situation radically. The requirements for human beings to live full, satisfying, and meaningful lives are not superior to the needs of animals to live full, satisfying, and natural lives, but neither are they of less value. What such writers as Kerasote, Stange, and Nelson both argue and *show* is that, for some people, living a life of deep reciprocity with the natural world around us is an existential necessity. This can take different forms, and Carol Adams's ways of relating to the world offer one possibility. But hunting, too, provides a kind of participation in the process of life, with its ebbs and flows, its intertwining of life and death that our society tends more and more to insulate us against. Even vegetarianism, as Kerasote argues, can become a way of denying the ways in which we are implicated in death and destruction. This is neither to accuse vegetarians of bad faith, although there are vegetarians who are in bad faith,[22] nor to absolve hunters in general from the charge of domination. What it does indicate is that there are a variety of ways to live respectfully and reverently in and as part of the natural world. The relational hunt is one of those ways.

RESPECT AND AFFIRMATION

Ethical theory often seems to move between two extremes. According to the first extreme, moral principles have a primarily negative function: they set limits for our actions and projects, giving shape to our lives by contouring them to the limits of the morally forbidden. Exemplary moral principles would be "Thou shalt not kill" and "Thou shalt not steal." In their secularized versions, such theories are constructed on the basis of a fundamental distinction between the forbidden and the permitted, where the permitted is anything that is not forbidden. Only in a situation in which all possible actions except one are forbidden can we say that a specific action is obligatory (an obligatory action being an action we are not permitted not to do). The more normal situation is one in which as long as one does not do anything forbidden, one can do as one chooses, picking from among the broad field of indifferently permissible actions as one pursues one's own happiness, satisfying one's needs and desires, and so on.

The other extreme is the idea that it is the task of an ethical theory to tell us what to do in any possible situation we may be in. Some versions of utilitarianism take this form, but Plato also suggests a version of it (*Euthyphro*, 6e). In the utilitarian version, in any situation we should add up the benefits and harms of the various possible actions, and perform that action that maximizes benefit over harm. Thus, in principle, if we had adequate information, there is one morally correct action that should be performed in a given situation. Only if there is a class of actions with, as far as we can tell, equivalent benefits and harms is there room for personal choice. Specifically moral decisions are made by calculation, not deliberation.

In Part II, I discuss Albert Schweitzer's rejection of this second conception of ethical theory. Schweitzer argues, correctly I think, that an exhaustive list of ethical rules that apply in principle in any situation robs us of moral responsibility, relieving us of the task of genuine moral deliberation and decision. But I also think that the first extreme is incomplete. To stay within Schweitzer's ethical theory, if we accept the principle of reverence for life then certain actions are simply forbidden. In particular, any action that destroys or harms life in any form is immoral if it is thoughtless or unnecessary. Here his ethical principle gives very precise guidance. But the principle of reverence for life does not begin to give this kind of precise guidance in all situations, nor does it aim to do so. Given the fact that life is a system of interdependence, that life lives from life, human life does require that other life be harmed and destroyed. But for Schweitzer this is neither simply forbidden nor a matter of indifferent permissibility. Rather, it raises the question how we are to live lives dedicated to reverence for life while living in the web of interdependence, and this is a *moral* question, even if moral principles cannot simply determine what we are to do. I argue in Part II that this question is posed correctly only when the concept of life is understood in terms of interdependence, and only when this system of interdependence and our position in it are *affirmed as being themselves good*. Only such affirmation allows us to live lives that are themselves affirmative—of ourselves and of the world in which we live.

Reverence for life as I have reinterpreted it is a principle that can guide us as we attempt to live lives that are worthy. More specifically, the attitude of reverence for life focuses our attention on how we can live with our actions if we are deeply mindful of self and world. The question is always whether we can really live with ourselves when we are at our most aware (cf. Plato, *Gorgias*, 482c; Arendt, 181): what kinds of lives we will have to lead if we fully recognize that other persons are entitled to our deepest responsiveness and what kinds of lives we will have to lead if we cultivate the attitude of deepest respect for the world and all that lives in it. Reverence for life gives shape to our lives, choices, and attitudes, even when it does not dictate them.

But as I have already suggested, this shaping can occur in two different ways. Some actions and practices, such as thoughtless and frivolous destruction, or actions that undermine the integrity of living systems, are simply wrong. Other actions and practices, including certain hunting practices, are a matter of personal responsibility and authenticity *within* the attitude of mindfulness, respect, or reverence. Here the basic moral principle does more than separate the forbidden from the indifferently permissible. It has a great deal to do with the *manner* in which we do those things that are in fact permissible in the sense that they are in the field of our own personal, subjective decision. Such decisions and the resulting practices are anything but morally indifferent. They require of us the deepest mindfulness and respect we can attain. It is noteworthy in this regard that serious publications devoted to hunting often show great respect—as indicated by publication—for those who once hunted, but now find that they cannot do it any more (cf., e.g., Wilder). The figure of the passionate hunter who hunts less and less, and finally, as he or she gets older, not at all, is a familiar one in hunting literature, but so is the figure of the hunter whose hunting gains in awareness and moral depth as he or she gains in years and experience.

Once one has accepted the basic moral principle, the task of moral deliberation truly *begins* as one is confronted with the task of living one's own individual life, of taking personal responsibility for one's decisions and practices. This is the task of living constantly mindful of one's basic orientation toward life expressed in basic principles or attitudes, without a set of rules that make one's decisions for one. This means that your decision, for example, to hunt, as long as it is consistent with or a personal expression of the basic attitude, can be correct *for you* without necessarily being the correct decision for me, even if we both adopt the same basic attitude. The moral life is a *way*, not the simple application of a set of moral rules. Many paths can diverge from one another within the common pursuit of this way.

Applying this general approach to environmental ethics, I contend that the background assumption for any environmental ethics has to be that the appropriation and assimilation of energy, nutrition, and other things that are required for our physical, mental, and cultural health, are *prima facie* permissible—indeed good. Our practices may be subject to limits of various kinds in the sense that they must be appropriate as expressions of the basic principle or attitude of respect for life in all its forms, but we are part of the biotic world, and there is nothing morally wrong with this status. It is a status that must be *affirmed* by any environmental ethic. Any recognition of inherent worth or intrinsic value outside such an affirmation remains abstract and unreal—both with regard to ourselves as biological beings who are moral agents and to the living world of which we are part. The task of environmental ethics is to help give shape to such practices.[25] Beyond the fact that we are part of and therefore engaged in the natural world, our engagement takes the form of actions and practices that are meaningful for us. No principle that tends to forbid such practices in general, or deny that they have affirmative meaning for us, no matter how qualified, is an acceptable starting point for environmental ethics. *And this claim is fully consistent with biocentric egalitarianism, properly understood.* It gives human beings no superior status; it is simply to acknowledge *and affirm* that we are participants, and potentially *respectful* participants, in the flux and flow of life, the web of interdependence.[26]

Only if we have already removed human beings from being an integral part of the natural world to begin with is action that is an expression of ourselves as members of the biotic community an affirmation of superiority and of our unique intrinsic value. As I noted in the Preface, this removal can go in either of two opposite directions. Traditional anthropocentrism removes us from the natural world and gives it to us as our instrument, to be used in any way we see fit. But it is also possible to remove human beings from the natural world in the opposite manner by positing a moral obligation that we minimize, to the greatest extent possible, actions that have an effect on the natural world. Here the ideal would be that human beings have no impact on nature and wild organisms at all, effectively removing human beings from nature. For the former, hunting is not a moral issue at all in any direct sense. For the latter, hunting is permissible only when it is absolutely required for survival, and, more importantly, hunting should not under any circumstances be given a positive meaning beyond that of serving survival.

The first step toward taking this dilemma by the horns is to take seriously the idea that nature, including the human species, is a web of interdependence. Murdy's position, when interpreted as I suggest above, is *biologically* anthropocentric—and because this anthropocentrism is biologically based, his concept of "anthropocentrism" is to that extent itself biocentric. Part of this anthropocentrism is the requirement that we, for purely self-serving, anthropocentric reasons, learn to live in a way that does not destroy the world we depend on. And his position is *culturally* one that requires that we lead our anthropocentric lives in terms of an awareness of the intrinsic value of the natural world of which we are a part. But this claim achieves its full resonance only when Murdy invokes a sense of our own *wholeness*, one that integrates self and world rather than separating them. Here the demand that we respect the inherent value of nature is not merely instrumental. The claim is rather that we will live better, richer, more satisfying, and morally superior lives when the ways in which we relate to the world around us are constitutive of our sense of self, when respect for the world of which we are a part is part of our respect for ourselves.

It is not the participatory hunter who is nihilistic, but the executives of the mining company who plan a gold mining operation on the headwaters of a river that runs into Yellowstone Park, the CEO of a wood products corporation who proclaims, "We log to infinity. Because we need it all. It's ours. It's out there, and we need it all. Now." (Harry Merlo of the Louisiana-Pacific in 1989; quoted by Chase, 305), the developer who markets a nature that he or she helps destroy by that very effort, the SUV ads that present the image of a canned wildness that is instantly accessed by the SUV of choice, the president who says that the current economic needs of U.S. business take precedence over the threats to other people, future generations, and biotic systems, and the vice president who says that energy conservation is at most a merely personal virtue, having no broader cultural significance, be it economic, political, or moral. We live in a culture that has lost touch with the fact that our sense of self is part and parcel with our sense of the meaning of the living world around us, or rather, we live in a culture that has compartmentalized what remains of that sense, limiting it to hobbies, vacations, and avocations, and especially to the marketable versions of these (cf. also David Strong, *Crazy Mountains: Learning from Wilderness to Weigh Technology*). The result is that while more and more people profess to love wild nature, and are willing to pay tax dollars to save "wilderness" (especially if it does not affect their ability to make money), these same people increasingly destroy the very thing they love by building everything from vacation cottages on five-acre "ranchettes" to trophy homes in the middle of what remains of habitat for animals such as grizzly bears, mountain lions, and elk. The nature they think they love and want to become nearer to quickly becomes an annoyance about which something needs to be done.

SPECIESISM REVISITED

Returning to the concept of "speciesism" from this perspective, I now make several points. First, everyone agrees that it is not speciesism to deny a moral status to animals that cannot meaningfully be attributed to them—thus, for example, the right to vote, but also the right to be treated with the specific respect due to persons as such. Second, Peter Singer is right in his claim that it is speciesist to argue that animal pain does not matter morally simply because it is animal, as opposed to human, pain. But, third, the fact that this pain is morally considerable does not mean that it should play the same role in human action as human pain, just as respect for animal lives cannot mean the same thing as respect for persons. The point here is not that animal pain is deserving of *less* consideration, but rather that *different* considerations come into play.[27] Just what those considerations are depends on the context. As Holmes Rolston III argues, "pain in ecosystems is instrumental pain, through which [wild animals] are naturally selected for a more satisfactory adaptive fit" (Rolston 1994, 112). Any assumption that it is a general and global human obligation to reduce the suffering of wild animals, as proposed, for example by Cleveland Amory, is a failure to respect precisely the wildness of the animals in question. To this extent, Rolston argues for a hands-off approach to wild animals when it is a matter of pain that is not caused by human beings.[28] But when Rolston calls for a hands-off approach to wild animals, he is far from making common cause with Tom Regan, Peter Singer, or Paul Taylor. And as Rolston sees, when we take as our point of departure the fact that human beings are part of the web of interdependence, and *affirm* human appropriation of and from the natural world, this does not mean that any and all appropriation is thereby justified or made permissible. That would be to fall back into traditional anthropocentrism

Rather, to take a first step, it means that as we appropriate the nourishment, energy, and materials we need we are under the moral constraint of avoiding cruelty, defined, following A. A. Luce, as the voluntary infliction of unnecessary pain and suffering. Economic efficiency, for example, is no justification, especially in well-fed communities. When "animals are treated like machines that convert fodder into flesh, and any innovation that results in a higher 'conversion ratio' is liable to be adopted" (Singer, 63), domestic animals are not being treated with an appropriate respect. They are indeed being treated *merely* as means.

But in different contexts there are different constraints. Viewed in terms of an affirmation of the existential fact of interdependence, the fact that a human being kills a nonhuman animal is not necessarily an expression of an attitude of superiority. Conscious, affirmative participation in the system of interdependence does not require the assumption of a moral or value hierarchy. If one is pursuing wild game, a participatory biocentrism both allows that pursuit and requires that it be done in ways that are consonant with, and thus respect and affirm, the wildness of the game animal. This will sound contradictory to many, but I have provided a framework that begins to explain why this is not so.[29] For example, elk evolved as the magnificent wild animals that they are under the predatory pressure of wolves, grizzly bears . . . and, for more than 10,000 years, human beings. Shooting (as opposed to hunting) elk in a hunting preserve, as is legal in some states (cf. Hering 2000b and many articles in *Bugle—Journal of Elk Country*, the magazine of the Rocky Mountain Elk Foundation), or over salt pits just outside the boundary of Yellowstone Park, which is illegal but has been happening for some years now (cf. Hering 2000a), not only fails to respect the wildness of the elk, in the hunting preserve it destroys that wildness. This is indeed to use the elk as a *mere* means, to regard it as a mere collection of physical properties, to be used and valued in any way that happens to suit some human being (e.g., as trophy). To use Carol Adams's language, to shoot an elk in a "hunting" preserve *is* to ontologize it as a trophy. But to hunt an elk one plans to eat is not in the same sense to "ontologize" the elk as "edible" (cf. Adams, 103). Elk *are* edible . . . to carnivores and omnivores, to hunters and scavengers. (This is the truth in Holmes Rolston III's insistence that eating is in the final analysis an event in nature.)

Elk have evolved as they have precisely because they both *must eat* and *are eaten*. That elk are prey animals is an evolutionary truth, written into their physical characteristics and their behavior. In an extreme sense (which should not be attributed to Adams), *not* to recognize elk as edible, as prey, is to "ontologize" them as something they are not—the legacy of Cleveland Amory. To hunt wild elk in a manner that participates in their wildness *is* an expression of respect for what they are. But such affirmation and appropriation places demands on the hunters, demands that the hunters have mastered the skills required, for example, and that the hunters not go beyond the limits of their skills. These demands are taken seriously in much of the hunting community, as one can see in the regular column titled "Situation Ethics" in *Bugle*. But beyond such questions of skill, such hunting places demands on the hunter's attitudes. Genuine respect for the animal hunted requires an expression of the hunter's own self-respect as a member of the biotic community.

Singer and Regan have their finger on something when they demand that human beings not treat animals, wild or domestic, as *mere* means to human ends. The fact that their interpretation of this duty tends to take human beings as moral agents out of nature shows, I have argued, that they have misinterpreted this requirement. When they reject human indifference to animal well-being, they are on solid ground. But since they (and this holds for Taylor even more, since he greatly broadens the range of ethical considerability) do not think in terms of an affirmative relation to our place in nature as beings dependent on appropriation, they *forbid* all appropriation to the greatest degree possible, rather than defining *appropriate* modes of appropriation.[30] These modes differ for domestic and wild animals. In neither case is cruelty morally permissible, but not all appropriation is domination.

Environmental ethics, when developed in terms of the participatory models suggested by Shepard, Lopez, and others, shows that individualism versus holism is a false alternative. The dilemma of an individualism that takes us out of nature and a holism that cannot account for the moral status of the whole (see Part III above) can then be taken by the horns. Just as I show above that inherent value cannot be separated from instrumental value when dealing with the relationship of any organism (including *Homo sapiens*) to its environment, there is a suppressed third in the supposed alternative of the individual living thing with its inherent worth and the system of interdependence of which that individual is necessarily a part. Human beings are both part of nature and thoroughly cultural. When we begin to reflect on the meaning of individual existence in a world of interdependence, and to ask what a proper relation to the natural world can be for an animal with the powers to transform and destroy what we have developed, then we must ask what a respectful relationship to wild nature can be in the modern, or postmodern, world.

Chapter 7

Toward a Philosophy of the Hunt

Since I have taken hunting as a test case both for my critiques of animal rights/animal liberation thought, Albert Schweitzer's ethics of reverence for life, and Paul Taylor's ethics of respect for nature, and for the first steps I take toward developing an account of an attitude of respect for nature that emphasizes and affirms that human beings are part of the world they live in, I end Part IV with a consideration of several topics that come up repeatedly in hunting literature.

HUNTING AND HUMAN SUPERIORITY—ORTEGA AND MEYERS

I have argued above that once it is recognized that life essentially involves webs of interdependence, it becomes clear that for one living creature to kill and eat another living creature has nothing to do with hierarchies of any kind. In anything that lives, intrinsic value and instrumental value are essentially intertwined with one another, and it follows that instrumental use of another organism does not necessarily involve any kind of intrinsic superiority of the eater over the eaten. I have argued for this above, mainly against those who condemn hunting because they think that it involves an assertion of human superiority and the exercise of domination. But the claim that the hunter is superior to the hunted is also part of some traditional philosophy of the hunt. José Ortega y Gasset's *Meditations on Hunting* is widely considered to be a classic work and among hunters is the most broadly influential philosophical discussion of hunting. Since Ortega argues that all hunting essentially involves a relation of a superior animal to an inferior animal, it calls for a direct discussion.

Ortega claims that "[hunting] is a relationship between two animals *which excludes an equality of vital level between the two*, and, of course, it excludes even more the possibility of an inferior animal's practicing it on a superior animal" (Ortega y Gasset, 61). Ortega thus assumes "the essential inequality between the prey and the hunter," insisting that "hunting is irremediably an activity from above to below" (*Ibid*). There are two points to be made here. First, the fact that ecologists speak of trophic levels (primary producers, primary consumers, secondary consumers, decomposers, etc., cf. Pianka, 339–340; Worster, 307–310) does not justify speaking of superior and inferior animals.[1] By the same token, we should be very careful when importing the notion of "superior" and "inferior" into relations of predator and prey, which need not even necessarily belong to different trophic levels in any simple sense. Members of a given species can be both predator and prey with respect to members of another species, depending on situation, life stage, and even simply luck. Once we look at hunting, whether natural predation or human hunting, against the background of the web of interdependence in which bacteria and parasites can kill human beings, using human bodies as nutrition and as media for their own propagation, it seems clear that the appropriation of that which sustains life has nothing to do with some kind of natural hierarchy of the superior and the inferior.

There is also no necessary presupposition of any hierarchy of value in human hunting relationships. Holmes Rolston III argues that there are such hierarchies of intrinsic value (cf. Rolston 1988, 223–225). I am not concerned to dispute this, only to insist that the moral permissibility of the hunting of animals by human beings does not rest on the claim that human beings as a species have higher (or more) intrinsic value than animals. It rests rather on a primal affirmation of the appropriative nature of life. In Part I of this book, I argue that human hunting and eating of other human beings is immoral not because human beings have a higher intrinsic value than animals, or because the hunter belongs to the same species as the hunted, but because as persons, as rational agents capable of asking for and giving justifications for our actions, we are responsible to other persons in a very specific sense. The result is that we have duties to persons that we cannot have toward creatures that are not persons. This does not, however, mean that we have no duties or obligations to animals. Indeed, I argue that we owe both wild and domestic animals and the natural world in general an appropriate respect. And I think that Rolston is right in arguing that our duties toward other persons do not always override our duties to the natural world (cf. Rolston 1998).

The mistake of much philosophical thinking about animals and about the broader world of living things is, as Rolston argues, to model our obligations toward animals on our obligations toward persons. This can be done by explicitly *extending* the moral status of humans to animals, then on to plants, species, ecosystems, and so on. On the basis of the work presented here, I conclude that there is one, but only one, element of truth in such thinking. This is the basic point that cruelty, the deliberate infliction of unnecessary pain, is always immoral. Indeed, this point needs to be made more strongly today than it was when Luce wrote. The system of industrial production of animals inflicts suffering in the name of economic efficiency that no society should tolerate. If the most effective way to convince a wide audience of this is to use extensionist arguments and rhetoric, then so be it. But the principle will not be understood correctly, and the distortions the extensionist argument brings with it avoided, until the principle is freed from extensionist rhetoric.

Beyond programs that are explicitly extensionist, theories of environmental ethics that are *modeled* on the ethics of interpersonal relations run grave risks. I suspect that it is such modeling that led Taylor to develop a concept of "respect for nature" poorly fitted to his own biocentric outlook on nature, putting an anthropomorphic relationship of "trust" at the center of human relationships with wild animals.[2]

But there is another issue of superiority in contemporary hunting, and Ortega comes close to putting his finger on it. Based on his view that "hunting is what an animal does to take possession, dead or alive, of some other being that belongs to a species basically inferior to its own" (Ortega y Gasset, 62), Ortega argues that "there is, then, in the hunt as a sport a supremely free renunciation by man of the supremacy of his humanity" (*Ibid*, 63). I have already argued that Ortega is wrong is his general claim that in all hunting there is a relation between the superior and the inferior in any sense beyond the fact that the hunter has to kill or take possession of the hunted in one way or another. But there is an important element of self-limitation and self-discipline in respectful hunting in our technological world. Ortega is certainly correct when he writes that the human hunter "could annihilate quickly and easily most animal species, or at least precisely those that he delights in hunting" (*Ibid*, 63). Both hunting and fishing magazines are full of advertisements for gadgets designed to make hunting easier and to assure "success" with minimal effort, knowledge, and skill. Ortega argues that the point of such limitations is that they increase the delight of the hunter, the "pleasure"

(*Ibid*). But while the delight and satisfaction that comes with mastering a difficult skill is undeniable and important, this is not the essence of the matter, since in this form it is irremediably anthropocentric. Alan Jones writes, "Any code that refers only to the hunter contains no teeth at all. If I'm acting a certain way only for my benefit, then there's nothing to keep me from acting another way. As long as the hunter is the only concern, then there is no possibility of a real ethic" (Jones, 67). The other that forms a pair with the solitary hunter is the natural process of life itself. It is respect for the wild animals we may at times hunt that leads to the requirement that our means of hunting them be appropriate. Limiting our technology is important because it is an expression of our respect for the animals hunted and for ourselves as hunters.

Stephen J. Meyers writes,

> *I think Gasset was right about the thing that constitutes sport: accepting limits, introducing the possibility of failure, knowing that the aesthetics of a hunt matter every bit as much and likely more than the count of dead animals at the end of the day. But this is not where the spirituality of a hunt resides, and I believe he had the matter of superiority all wrong* (Meyers).

Meyers runs through a series of arguments put forward in defense of hunting: the utilitarian argument that hunters are the source of dollars that protect and improve habitat, the argument that hunting is a tool of population control, and the argument that it is often hunters who cherish and protect what is left of wildness in our world.

> *It is clearly not just hunters and fishermen who speak for the earth and for wildness. It is entirely possible that some would continue to argue for wildness even if no man or woman were ever to hunt again. Their argument would lack something crucial, a degree of familiarity that can be gained in no other way, but a few would still struggle to save the wild. It is not because of the need to save the world that we need hunting. Glaciers and epochs and geological time will take care of that very nicely. We need hunting to save ourselves. And this, I believe, is the real meat of the argument* (Meyers).

The claim "we need hunting to save ourselves" will be offensive to many nonhunters in any number of ways. At first glance it seems to be deeply self-serving and to turn wild animals into instruments of our own satisfaction, this time under the pompous title of our own salvation. Meyers has argued that in a culture in which we are increasingly alienated from the appropriative processes of life,[3] hunting is one of the few completely *honest* things we can do.[4] In hunting and fishing, we relate to wildness *in its own terms*, which are those of interdependence and direct appropriation, to our own wildness as parts of the web of interdependence and appropriation, and to the wildness of the prey that is part of the same system. We need this "to save ourselves" *as parts of, citizens of*, the world, living as we do in a world that increasingly isolates us from this web of interdependence and appropriation. Meyers's basic point is personal—*he* needs hunting to save *him*self—but his point is broader than that. Alan Jones writes, "In the same way that social ethics is designed to protect society, a process ethic must be designed to protect the natural world. And if hunting is part of our larger, human relationship to the process, this ethics must be a subdivision of a larger, all-encompassing ethic" (Jones, 66). A doctrine of personal virtue, though important, is by itself inadequate here. Thoreau did not write, "In wildness is the preservation of humanity," but rather "In Wildness is the preservation of the World" (Thoreau, 672). We cannot save ourselves without saving the world, and we cannot save the world without saving ourselves. Individual integrity is crucial, but equally important is the world to which we relate in our practices. The way the world makes itself known in our practices informs both our ethics and the kinds of persons we become as we act.

Not everyone needs to hunt as a matter of personal authenticity, but in our cultural values we need to acknowledge and affirm the place of respectful hunting as one aspect, but a crucial and essential one, of the way we take our place in the broader world of which we are a part. Just as it is important to many people to know that wolves once again roam the Yellowstone system, even if they know that they will never see one, I think that it is important that we be a culture in which respectful hunting is practiced and taught—important even to those of us who do not hunt.

For this reason, it is good for us, all of us, that in many states there are public lands that are open to hunting and fishing and that hunting is not the class-bound practice that it is in much of Europe. By the same token, I think that it is not good for the culture of the state of Texas that there is almost no publicly-accessible hunting land in the state. It is good for our culture that there are people like

Richard Nelson, Ted Kerasote, Mary Zeiss Strange, Steven J. Meyers, and many others, who both hunt and reflect deeply on their hunting, and then share both their experience and their reflection. Their writing is important for the educational effect it can have both on other hunters and on those of us who either cannot or choose not to hunt, but find it important to think about these matters and to know that people like this are hunting in such a deeply respectful manner.

GUILT AND REMORSE

Hunters often write of the "remorse" or "guilt" that they feel after killing an animal. In an earlier chapter I quote Thomas McGuane on the remorse anyone "who loves to hunt" feels at the death of the quarry. Buddy Levy writes of feeling "exhilarated and sad at the same time," and of experiencing "a twinge of guilt in my heart" (Levy, 5–6) after killing his first chukar partridge. In his *Meditations on Hunting*, José Ortega y Gasset writes,

> *Every good hunter is uneasy in the depths of his conscience when faced with the death he is about to inflict on the enchanting animal. He does not have the final and firm conviction that his conduct is correct. But neither, it should be understood, is he certain of the opposite. Finding himself in an ambivalent situation which he has often wanted to clear up, he thinks about this issue without ever obtaining the sought-after evidence* (Ortega y Gassett, 98).

Here is one final example. George Bird Evans writes:

> *I live more richly, more miserably, and more wholly during the grouse season than at any other time of year. There is a mix of pleasure and hurt; for we who shoot know a guilt from knowledge that we kill and yet enjoy it* (G. B. Evans 1982, 9).
>
> *If shooting a ruffed grouse is no more than transforming a live bird into a dead one, it classes as a sin* (*Ibid*, 71).
>
> *After shooting a grouse, a short period of remorse sets in with regret and empathy for the thing you've killed . . . It might be called a not very sincere emotion, for to renounce this experience with a grouse and a dog is something I couldn't*

do. But it is a deep emotion even so, and it surfaces each time a grouse is shot, together with the realization that the gunner's act of possessing the wild is an intrusion (*Ibid,* 83).

Evans is especially interesting here, because he uses three different terms: "guilt," "sin," and "remorse." I interpret his use of "sin" (in this specific context) and "remorse" as an expression of his respect for nature, and specifically for the grouse he hunts. He appropriately feels remorse at the death of the beautiful bird in spite of the moral appropriateness, the pleasure, and the beauty that is part of hunting birds with shotgun and bird dogs. He properly calls the mere killing of a ruffed grouse, an act that is not part of the practice of hunting a beautiful (and delicious) bird with a well-trained dog and an appropriate gun, along with deep appreciation of the beauty of the bird and a deep awareness of and concern for its ecology, a "sin." But his use of "guilt," or his feeling of guilt, betrays a remnant of an attitude more akin to those of Schweitzer and Taylor. Even Ted Kerasote cannot avoid this language. "When people evolved a conscience, understanding for the first time the death that ran the world, they could no longer be one with all its other creatures—living in innocence of our mutual dependence, and killing without remorse. Call this our first guilt, or our first responsibility" (Kerasote, 225). Kerasote's use of the word "guilt," like use of the word "sin" in the similar contexts, evokes an imagined life without guilt, without sin. To become aware of weighty *responsibility* is one thing, but to say that we are *guilty* when we take life to feed life means that such a purity would be better, that we encounter here a fundamental defect in the order of our world. This is what the attitude of respect for nature should reject. It is a permanent temptation in our alienated culture, but one we should resist.

A deep sense of responsibility and sadness over the taking of a life that we value both for its own sake and for the role it can play in our lives is constitutive of the attitude of respect for nature. Respect demands and is the fruit of mindfulness. Schweitzer says that such mindfulness immediately makes us aware of our inevitable and ineradicable guilt. I argue that this is the result of a flawed mysticism. The fact that even our most deeply felt actions are many-sided and shot through with ambiguity need not be interpreted as a form of original sin. It does mean that mindfulness has to be cultivated.

Mary Zeiss Stange notes that when she began hunting, she refused to "start small, shooting birds or small game, working up to larger prey" (Stange, 17). She resisted the implication of a hierarchy in which the life of a rabbit matters less than that of an elk. "I cannot watch the death-arrested flight of a duck or a pheasant without an inner pang. Should the day come when I could, I would never shoulder a shotgun again" (*Ibid*). In the same vein, Kerasote cultivates practices in which the elk he is about to shoot is individualized in the moment of apology before the shot. George Bird Evans writes that each bird that is not eaten soon after being killed, when the memory and appreciation of that place, that bird, that point, that shot, and that retrieve, are still close, is frozen with a label detailing the circumstances of its death, so that when it is thawed and eaten, that past, which includes that particular bird and its death, can become present.

Richard Nelson recounts two experiences of powerful intimacy with two different deer. In the first, he reports being approached so closely by a deer unable to catch his scent that he is able to touch it (Nelson 1991, 274–275). In the other, he has the extremely rare experience of watching a doe give birth from a very close distance (Nelson 1997, 341–352). Each experience is one of great power, wonder, and intimacy, a moment in which an unusual contact with a being that is very other is achieved. But they do not put an end to Nelson's hunting. "I am not a guiltless [NB!] hunter, but neither do I hunt without joy. What fills me now is an incongruous mix of grief and satisfaction, excitement and calm, humility and pride. And the recognition that death is the rain that fills the river of life inside us all" (Nelson 1997, 340). As I noted in Part I, when Nelson speaks of living with the Inupiaq Eskimos and taking up hunting for the first time, he does not speak of it in terms of "pleasure," as in "hunting or killing for pleasure." He speaks rather of "the deep sense of satisfaction I discovered in that process" of finding and killing animals, butchering them, and making them food (Nelson 1994, 82). What he discovered was not the "fun" of hunting and killing animals, but the deep meaning that it can have when it is an integral part of the way one lives.

But there can be different experiences, with different outcomes, for different people in different contexts. In *Making Game: An Essay on Woodcock*, Guy de la Valdéne describes the rare experience of seeing a well-camouflaged woodcock on the ground.

> *Over the years, I have stared at a million leaves looking for woodcock and have been rewarded with a million puzzles, but this time the bird jumped out of the enigma a scant yard in front of [his dog's] nose. He was hunkered on the ground facing away and watching me out of his famous eyes. There ensued a time warp of sorts, during which the only suggestion of motion was the dog's shivering flanks* (de la Valdéne 1990, 127).

When the woodcock finally flies, de la Valdéne does not shoot, and it is his hunting partner who kills it. But the experience stays with him.

> *I often think about that bird on the ground and am glad the occurrence is rare. It would be impossible for me to kill woodcock after seeing them bundled up like elves in the leaves. I punish myself enough by canting cripples towards the sun and looking into their pupils before crushing their heads with my thumb. I don't know what I am looking for, but what I see is a deep blue reflection of my face. Killing is already too serious and precarious an affair for me to be that calculating an executioner. Birds in flight are targets, but on the ground they embody a form of latent freedom we all long for, the power of escape* (*Ibid*, 128).

One might think that this is a version of what Adams calls "the absent referent," now not as a function of language, but as a function of the contours of specific kinds of practices, different kinds of experiences. George Bird Evans writes, "Upland shooting is at its highest level when the bird-dog-gun triad is balanced by a gun and a dog worthy of the bird" (G. B. Evans 1971, title of photograph facing page 113).[5] The problem is that, as de la Valdéne notes, at the moment of flight the bird becomes a target. It is perhaps partly to counterbalance this that Evans so carefully cultivates an appreciation of the beauty of the bird both immediately upon its retrieval and later when it is eaten, but this functionalizing seems ineradicable. The bird on the ground is, after all, the same bird as the one in the air, but the exercise of a skill and the experience of success can dominate the moment. This is also the reason why Evans finds the aggressive language of many hunters inconsistent with an attitude of "respect for your game," and thus carefully avoids language expressive of "hostility between the gunner and his game" (e.g., "bust a grouse" and "clobber a pheasant") and language more appropriate to the shooting of clay targets (*Ibid*, 10).

But there is another way of understanding what is happening at the moment the grouse becomes a target. Alan Jones makes an important distinction between the *project* of hunting, the goal of which is to kill the prey, and the *purpose* of hunting, which is not simply a matter of killing an animal (A. Jones, 55). The purpose of hunting as Jones understands and practices it is to participate actively and consciously in the process of life (Jones's equivalent to Paul Taylor's web of interdependence). This distinction enables Jones to argue that contemporary hunting is sport on the level of project, but not on the level of process. "The project, the sport, allows us to orient ourselves within the natural order. But the order that produced the sport requires our first allegiance. And this is the process" (*Ibid*, 56). The mistake is to lose this sense of balance. "The animal must never be a means to an end. If you want to acquire a trophy so badly that you're willing to do anything to get it—if you let the project take precedence over the process—then the animal has been reduced to an object, a means to an end" (*Ibid*, 57). But when the process is allowed its orienting priority, "Hunting is the deobjectification of the animal. The act of killing is the final stage of deobjectification. The grouse swinging at my waist *is* the world. Tonight, I'll go to bed feeling that I have been a part of something complete" (*Ibid*, 61; cf. also 24–27, 39).

Jones's concept of what he calls "deobjectification" is based on a phenomenological analysis of the act of hunting itself. Many writers on hunting emphasize the aspect of skill in tracking, finding, stalking, shooting, and so on, and find an important part of the value of hunting in the acquisition and exercise of these skills. One important aspect of this exercise of skill is that it puts the individual in touch with something deeper than the conscious, thinking, and deliberating self. But Jones detects yet a deeper level at work.

> *The entire relationship exists on a level beyond language. By hunting, your life moves away from subjects and objects. You're part of a relationship larger than either of the two animals involved. You're no longer precisely yourself. Hunting requires you to lose yourself in the act.*
>
> *Not that this is any great revelation. Most acts of skill require a loss of self. If you stop and think how you're going to hit the tennis ball you've already lost the point. But what is a*

> *revelation is the absorption of the act by the process. You lose yourself in the act and then the act is lost in the natural order. How wonderful! (Ibid, 40).*

The key is not found in the skillful act itself, but in the way this kind of skillful act relates to and is a part of the process of life itself. Both are required for the experience that "the grouse swinging at my waist *is* the world." The bird becomes a target as the hunter is absorbed in the act of shooting, but the entire event—search, flight, shot, kill, eating—is "lost in the natural order," even though—or rather because—the hunter brings a great deal of culture, of hunting culture, to the scene. Pete Dunne evokes this entire complex in the moment that "lies between the shot and the echo" (Dunne, 31–32). For Dunne, the distinction between the hunter and the non-hunter amounts to: "I have stood in the moment before the echo. They have not" (*Ibid*, 33).

Alan Jones's understanding of what it means to treat an animal as a [mere] means to an end should be compared with the understanding found in Tom Regan's animal rights theory. Jones presupposes that our lives are intertwined with those of animals, and his insistence is that we do not turn animals into mere things, objects to be manipulated according to our changing values and whims. Regan takes us out of the world, attempting to undo the intertwining to the greatest extent possible. Similarly, for Jones an authentic attitude of respect for wild animals is an expression not only of the nature of the animals, but equally of ourselves as parts of the same process. In order to respect wild animals, one must accept not only them but also oneself as part of the existing order. Any other attitude inevitably objectifies the animal (cf. *Ibid*, 25–27, 50). "If hunting is your project, you have stopped observing nature and have begun participating in it. Rather than attempting to impose a new order on the process, you have accepted the existing order" (*Ibid*, 26). This understanding of "respect" should also be compared with Paul Taylor's attitude of respect for nature, the tendency of which is to take the human moral agent out of the process. From Jones's perspective, Taylor's attitude is a product of objectification.

But even this does not necessarily banish the tension for a given individual, and this is the realm of authentic subjectivity. *Making Game* ends with Guy de la Valdéne writing, "I will hunt him [the woodcock] because it gives me intense pleasure to do so, and if I kill him I will eat him and love him only more. When the riddle has run its course, I will stop hunting" (de la Valdéne 1990, 161). But in the "Introduction," written more that a year after completing the manuscript of the book itself, he writes, "I realize now that I had been in love with a bird, a beautiful and painfully dumb little bird, predictable and thus vulnerable in every respect. Therefore, I don't like to shoot them anymore" (*Ibid*, x). This is written in the context of dwindling populations of woodcock, a function of habitat loss and "an explosion of yuppies (guppies) bearing guns in the 1990s" (*Ibid*, ix), but it is hard not to hear the echo of his experience seeing the woodcock on the ground, feeling those "famous eyes" looking at him, and identifying with the "latent freedom we all long for, the power of escape." Hunting is an activity that takes place on the edge, indeed on many edges.

HUNTING VERSUS PHOTOGRAPHY, BIRD WATCHING, BERRY PICKING, ETC.

I noted above that in an attempt to describe what is so powerful and unique about the experience of hunting, hunters have often felt the need to devalue other activities that relate to nature such as hiking and photography. Antihunters answer in kind, arguing that berry picking or gathering mushrooms is just as much a participation in natural processes, the only difference being that such activities do not involve killing sentient animals. This allows the antihunters to focus on killing, leaving aside the issue of the meaning of the project of hunting. The hunters, in turn, focus on the act of hunting and consider the act of killing only from the point of view of the hunt as a project. Some justify the importance of the hunt by claiming that other practices do not measure up to the profound experience of hunting.

Ortega y Gasset offers the classic, and most entertaining, attack on wildlife photography.

> *The English have initiated a form of hunting in which all these conflicts of conscience are cleverly eluded: it is a matter of having the hunt end, not with the capture or death of the animal, but rather with taking the game's picture. What a refinement! Don't you think so?* (Ortega y Gasset, 92)

> *Photographic hunting is a mannerism and not a refinement; it is an ethical mandarinism no less deplorable than the intellectual pose of the other mandarins . . . The mannerism consists in treating the beast as a complete equal, and it seems to me more authentically refined and more genuine to accept the inevitable inequality which regulates and stylizes the perennial fact of hunting as sport* (*Ibid,* 94–95).

Beyond the polemics, Ortega's position here is based on his claim that hunting is essentially a relationship between a superior and an inferior species. I have argued above that this is untenable, and distorts rather than clarifies the meaning of human hunting.

A more serious statement of the claim that hunting is uniquely superior to other ways of relating to the natural world is found in Allen Jones. Jones argues, "Hunting is the original project . . . Hunting is our only project that manages to deobjectify nature, to allow participation in it as a member" (A. Jones, 27). If this is the case, then it follows that other projects, all other projects, objectify the world. (It is not clear whether Jones would allow berry picking and mushroom gathering to qualify as forms of hunting, but he focuses on other activities.)

> *If you're hiking, nature is something only to walk through and observe. The requirements of the project do not require interaction with the natural world in any meaningful way. You're walking* through *it, not* in *it. For the hiker, the world is . . . scenery . . .*
>
> *And if hiking is in bad faith, what about other human activities. Water skiing? Gold mining? Photography? If you're taking a picture, the animal and the features of the landscape are pure objects. Within the frame of the lens, they are no longer moving, no longer active, no longer alive; instead, they've been caught as arrested reflections. Nature photography drags from the world a meaning that requires you to stand with your hands behind your back, watching an instant of nature and taking it for reality* (*Ibid,* 25–26).

Jones's rant against photography, putting it in the same category as gold mining, should be read in the same spirit as one can read Joy Williams's rant against hunting: there is a lot of truth in the words even as they completely miss the heart of the phenomenon at its most authentic. Jones's mocking description of "wildlife" photographers flocking into Yellowstone Park in early fall to photograph "these domesticated elk" (*Ibid*, 26) is right on target, but to identify what goes on there with "photography," or even with "wildlife photography," is an act of willful ignorance. The fact, noted above, that many passionate hunters are photographers, bird watchers, hikers (cf. especially David Petersen's *Elkheart*, in which hunting plays very little role, though Petersen is passionate about hunting) is enough to indicate that Jones's rhetoric is taking over.

It is true that for many hunters there is nothing else quite like the intensity of the hunt. This is an experiential truth, but the question is whether the experience gives us access to a deeper philosophical insight. Thomas McGuane suggests that it does.

> *Whatever drives some people to hunt lies in a great skein of elements that are beyond selective human control . . . Fishing, of course, is a part of hunting and anyone who has not picked up its instruments and gone forth to feel the transmutation of the country before them has experienced a profound omission. It is what Orwell called a hole in the light* (McGuane 2000, 16).

This is a beautiful evocation of the experience of the hunter and angler, but McGuane's own text undermines the "anyone." *Some* people are driven to hunt . . . and others are not. Some people are drawn to trout or elk, and others are not. For some the instruments that give access most intensely are fly rod, rifle, or bow and arrow, and for others this is utterly incomprehensible. For some, a pair of binoculars and a few minutes of bird watching are a hole in the light, whereas for others it is just a meaningless looking around for bird feathers. What for one person comes immediately and naturally is for another a learned pleasure, and for yet another forever remains foreign.

In other words, I think that there is something fishy about these arguments, regardless of the individual authenticity that comes to expression in them. The source of the error is most clear in Allen Jones, largely because his analytic descriptions of hunting are so acute that when he makes a mistake it is easier to spot. Jones's distinction between the project and the purpose of hunting is, I have suggested, well-drawn and important, and his analysis of the deobjectification that is part of the authentic hunting experience is deep.

But Jones pushes this analysis one step too far when he writes, "If we could ever fully return ourselves to the process, we would have no need for an ethic since we would have no ability to choose our actions" (Jones, 108). Note that this presupposes that as cultural beings we are not and cannot be full members of the process. This leads to his suggestion, "For the complete hunter . . . his acts would exist *without intention*" (*Ibid*, 114). This is romanticism, the ideal of returning to the origin and merging with it.[6] Indeed, it is contradictory: the idea of "*returning ourselves* to the process" is the project of becoming beings without projects, the intention of living "without intention." This is a false ideal for a cultural being, and Jones's own analyses tell a different tale. Elsewhere, Jones understands that human hunting, which necessarily moves within the ethical dimension, requires consciousness and conscious intention. This is not a lack of completeness, but part of the essence of hunting as a human project. It is this false ideal that is the source of Jones's claim that only hunting overcomes—however imperfectly, since he claims that the human hunter is never the "complete hunter"—our alienation from and objectification of the natural world (*Ibid*, 41).

"ZAP-AND-RELEASE" VERSUS FAIR CHASE

These edges have moved some hunters to consider an analogue to catch and release fishing in the realm of hunting in the narrower, everyday sense of the term. George Bird Evans writes,

> *If I could shoot a game bird and still not hurt it, the way I can take a trout on a fly and release it, I doubt if I would kill another one. This is a strange statement coming from a man whose life is dedicated to shooting and gun dogs. For me, there is almost no moment more sublime than when I pull the trigger and see a grouse fall. Yet, as the bird is retrieved I feel a sense of remorse for taking a courageous life. About the time I passed fifty I noticed this conflict becoming more pronounced* (G. B. Evans 1971, 7).

This is clearly not a response to dwindling populations of game animals, although Evans is very concerned about that and a severe critic of what he considers the naïveté and self-satisfied know-nothingism of many game managers (cf. G. B. Evans 1982, 113–176). Evans is trying to imagine an idealized version of the practice he loves, hunting grouse with magnificent bird dogs and a fine gun. And he is not alone. Frank Jezioro writes, "Heaven to the bird hunter would be an old road through an eternity of grouse cover that he and his never-aging dog could walk through while he carried a fine double shotgun. The dog would point every bird, and the hunter would kill it, only to hold it for a moments [sic] admiration before releasing it to fly again and again" (Jezioro, 17). This would be hunting without killing, or killing without death, hunting without remorse. But it is no accident that this is the dream of a timeless present, of event without consequence, of repetition that is ever the same, of action without drama, of life without the process. Add real consciousness and memory to the hunter in this fantasy and it becomes an image not of hunter's heaven, but of a hell of empty repetition, a bloodless (and foodless) fake of a world, a world without process, without interdependence, without life. A practice that does not allow for failure is not a meaningful practice, and a life in which death plays no role is not alive.

Some have played with the notion of the hunt without the kill more seriously. George Bird Evans considered mounting a miniature camera on his gun, complete with crosshairs so that the "success" of the shot could be determined (G. B. Evans 1971, 7). Similarly, it has been proposed that elk be "hunted" (perhaps the more appropriate word would be "stalked") with paint-ball guns. The criterion of a "successful" stalk is the splash of paint on the elk, which can run away and continue to live. Such a practice would be similar to wildlife photography without telephoto lenses, but with an emphasis on something like an experience of counting coup. The idea of elk walking around with splotches of paint on them strikes some hunters as truly disrespectful, a toying with the wildness of the elk, playing with its vulnerability rather than engaging it. A high-tech version would use laser guns, and could be adapted to bird hunting as well. Edward Hoagland proposed (and predicted) something along these lines in his review of Howell Raines's *Fly Fishing Through the Midlife Crisis* for the *New York Times* (cf. R. Jones, 165–166). Robert F. Jones, with whom Hoagland discussed the idea before writing his review, considers this proposal, which he calls "zap-and-release," rejecting it because it turns hunting into a game.

> *The hunt is more than a gallery game. At its best (and at the risk of sounding absurdly uncool) let me say that I feel it can be a sacrament . . . And each time we partake of the flesh of the birds we've killed, we become one with their essence.*
>
> *Hunting is incomplete without death.*
>
> *That's why it can never be a mere gallery game* (R. Jones, 168–169).

And yet Jones admits to Hoagland that most of his fishing is catch and release. Why this discrepancy? It may be that this feeling will fade, that in the future "hunters" with laser guns will compare notes at the end of the day. That would be no stranger, and certainly more honest, than proudly displaying the mounted head of a "shooter elk," bred and raised on a closed ranch, and offered to anyone willing to pay the price, results guaranteed (cf. Hal Herring's article in *Harper's*).

For George Bird Evans, the senselessness of the gun-mounted camera was pointed out by his wife Kay. "Kay points out that shooting is a part of me, and that the dogs deserve to have birds shot over them and I know she's right" (G. B. Evans 1971, 7). To dilute the practice of hunting would remove something essential for both the hunter and the dogs,[7] so Evans remains caught in the dilemma.

> *How then, can you love a bird and kill it and still feel decent? I think the answer is, to be worthy of your game. Which boils down to a gentleman's agreement between you and the bird, never forgetting that it is the bird that has everything to lose. It consists of things you feel and do, not because someone is looking or because the law says you may or must not, but because you feel that this is the honorable way to do it* (*Ibid*, 8).

This is something like Barry Lopez's notion of a "contract" between hunters and hunted, "derived from a sense of mutual obligation and courtesy" (Lopez 1991, 381). When Evans speaks of a "gentleman's agreement," he is clearly using a metaphor, but he is using it seriously. The "agreement" is, of course, in the first instance with himself, but it has profound implications for the way he relates to the world, for his "congruency" (Lopez 1987, 297) with the world he is part of, and for the obligations incurred in the thoughtful exercise of his practice. In this sense, it is not only an "agreement" with himself, but with the world, with grouse collectively, and with individual grouse. It involves, to give just one example on each level, commitments about how humans should use and live on the land they share with birds such as grouse, the restrictions that hunters should put on themselves to protect grouse populations, and the responsibility the hunter bears to kill cleanly and to find and kill cripples. Beyond his professional work, George Bird Evans's life was a life devoted to play. He and his wife knew how to play with each other and with their dogs, and much of this play was, in Stephen Bodio's words, "elaborate ways of playing with your food and with the universe, ways that also give you windows into the lives of things as alien as insects [for Evans it would be grouse]. . . . or into the minds of canine . . . partners" (Bodio 1997, 230). But Evans also understood the seriousness of the play, and the responsibilities it involves.

Joy Williams targets such talk as that of the "gentleman's agreement" when she writes that the way hunters talk often implies "a balanced jolly game of mutual satisfaction between the hunter and the hunted—*Bam, bam, bam, I get to shoot you and you get to be dead*" (Williams, 252). To a modern sensibility, it goes without saying that prey animals do not "agree" to be hunted or killed, and as Theodore Vitali argues, talk of "fair chase" cannot invoke what he calls "the right to fairness in competition, that is, the right to be informed about and to consent to participation in the contest" (Vitali 1996, Part I, 21). This kind of fairness is specific to the ethics of interpersonal action, since only persons can be informed and give consent, and for this reason cannot be applied to person-animal interactions. As Vitali puts it, hunting is not a contest between hunter and hunted, and if it were, "hunting would be inherently unfair and immoral" (*Ibid*, 21–22; cf. A. Jones, 56).[8]

The larger point is that the obligations we assume when we hunt with an attitude of deep respect are not modeled on the obligations we owe to other persons. As I argue in Part I, when one attempts to extend inter-person ethics to animals, the counterintuitive result is that our obligations to animals are much stricter than our obligations to persons: we are obligated simply to leave them alone, stay out of their lives and their world, in spite of the fact that we, persons and animals, are all interdependent parts of the one biotic world. When Barry Lopez writes, "We will never find a way home until we find a way to look the caribou, the salmon, the lynx, and the white-throated sparrow in the face, without guile, with no plan of betrayal" (Lopez 1989, 388), his target is, for example, the logging, oil, mining, industrial agriculture, and other corporate interests for whom these wild animals are simple hindrances "in the way of our agriculture, our husbandry, and our science" (*Ibid*). These, along with the scientific revolution, Lopez thinks have led to "a determined degradation of the value of animal life" (*Ibid*, 382). But hunting a wild life in a way that respects both its wildness and the integrity of the system of which it is a part, taking a wild life in order to feed both our bodies and our culture, both our physical and spiritual lives, allows us to look the caribou or salmon we kill "in the face," to have respect both for the animal and for ourselves. Hunting often involves deception, but not necessarily betrayal—betrayal of the process, the web of interdependence and our respect for it, that is.

In the following passage (recounted in the third person, though Kerasote is the hunter), Ted Kerasote meditates on what he calls the "unresolvable unfairness" involved in hunting—something that is ultimately grounded in the "unfairness" of life itself.

> *Raising the rifle, he said to the elk, "Thank you. I'm sorry." Still he waited, the rifle raised but not pointed, for though he could lose himself in the hunting, he had never been able to stop thinking about its results—that this living creature before him would no longer live, so that he could eat. And he didn't know how to escape this unresolvable unfairness, but he knew he would rather be caught up in this quandary with creatures close like this rather than with distant, unknown ones, whose deaths he could not own directly. Not that he consciously thought all this; he knew it by his hesitation* (Kerasote 1999, 78–80).

This is an existential choice, made reflectively, mindfully, and with great clarity. It is a version of what Barry Lopez calls "leaning into the light" (Lopez 1987, 413).

Theodore Vitali argues that the practice of "fair chase" concerns the virtue of the hunter, not the fairness of a contest between hunter and prey. This virtue has two dimensions. The practices of the hunter should both protect the environment and fulfill "the moral obligation on the part of each individual to act in virtuous ways for the sake of personal excellence" (Vitali 1996, 22). This excellence requires, as Ortega y Gasset noted, "a supremely free renunciation by man of the supremacy of his humanity" (Ortega y Gasset, 63), but as I argue above, this "supremacy" is not some kind of superiority of vital level or intrinsic value, but technical superiority. Vitali goes beyond Ortega in his claim that the virtue involved in the practice of fair chase is not merely technological restraint. He writes, "Hunting, as distinct from other forms of killing, requires an explicit intention to enter into the predator-prey relationship for the purpose of experiencing fully all that this relationship entails" (Vitali 1996, 23). Such experience is, Vitali argues, something of genuine value, such that the hunter who enters this experience properly becomes a better, more virtuous person. By participating on this most intimate level in the intertwining of life and death, "the hunter often achieves, even if unintentionally and by accident, a kind of basic insight or wisdom into the natural continuum of life and death" (*Ibid*, 24). Doing so requires not only self-restraint, but also the cultivation of a proper frame of mind and a proper practice. Fair chase thus requires the mindfulness that comes from respect. It is not a "sport" in the sense of a diversion (cf. Ortega y Gasset, 29–30; Kerasote 1999, 82).

My claim is that the strength of Vitali's argument lies in the fact that the two aspects of the hunter's virtue—the protection of the environment and the enhancement of personal excellence—cannot be two distinct considerations that are simply tacked onto one another, as if protection of the environment were the price the hunter has to pay for the right to hunt. Otherwise his position is simply another form of traditional anthropocentrism, now with an emphasis on stewardship. For a truly biocentric egalitarian view, the personal excellence involved is not that of the transcendent human being, the human being apart, using the natural world as an instrument for self-perfection, virtue, and wisdom. It is rather the excellence of the self that experiences itself as part of the world, the self for whom self-respect and respect for the world are two sides of one coin. There is no true excellence of self without recognition of and participation in the excellence of the world, and no genuine respect for nature without this same recognition and participation. This can be called "wholeness" (Murdy), "identification" (Naess), or "the ecological self" (Shepard). For each, our own excellence and virtue will be found in relating to and participating in the biotic world as that which nourishes both body and spirit, not in dominating that world. Part of the quest for genuine wholeness is learning to be a proper part.

RESPECT AS GRATITUDE

I argue that a crucial element of this virtue is an attitude of gratitude and thankfulness. While all appropriation is biologically self-serving and in that sense one-sided, genuine gratitude has the capacity to restore reciprocity. Barry Lopez writes, "For a relationship with landscape to be lasting, it must be reciprocal. At the level at which the land supplies our food, this is not difficult to comprehend, and the mutuality is often recalled in grace at meals" (Lopez 1987, 404). Gratitude is part of, and a condition for, genuine stewardship, especially for that paradoxical stewardship of the wild that requires us to protect it without managing it and to participate in it without destroying it, partaking of both the wildness without and the wildness within.

The words "sacramental" and "sacrament" have appeared in several of the texts on hunting I have quoted in this book (cf. R. Jones, 168–169; McGuane 1982, 236; also Rolston 1988, 91; Clifton, 148–149; Snyder, 19, 184). In the quote from Thomas McGuane (cf. page 19 above), he speaks of "a world in which a sacramental portion of food can be taken in an old way—hunting, fishing, farming, and gathering" as a good thing, as crucial to "societal sanity." If what is at sake is merely a certain vision of what "societal sanity" consists in, McGuane would be open to the charge of straightforward anthropocentrism: animals would be mere instruments in the human quest for societal sanity. But his use of the term "sacramental" moves in a very different direction, invoking images of the presence of and participation in the divine.

The language of the sacrament has a long and honorable tradition in American nature writing. In *A Week on the Concord and Merrimac Rivers*, Henry David Thoreau describes "an old brown-coated man who was the Walton of this stream." "His fishing was not a sport, nor solely a means of subsistence, but a sort of solemn sacrament and withdrawal from the world, just as the aged read their Bibles" (Thoreau, 65, 66). This language is fitting for Thoreau's transcendentalism. Speaking of another angler, Thoreau writes, "Thus, by one bait or another, Nature allures inhabitants into all her recesses" (*Ibid*, 64). But he also thinks of angling as a stage to be transcended. "This man is still a fisher, and belongs to an era in which I myself have lived" (*Ibid*).

But can such language still be taken seriously? As my colleague Larry May asks, Is it a metaphor? And if so, what is its cash value? I think that there are many ways of understanding this language, some of them more metaphorical, some of them less. None of them deserve to be taken with anything less than genuine seriousness. The metaphor, if that is what it is, is anything but "mere." Interestingly, a document entitled "The Columbia River Watershed: Realities and Possibilities—A Reflection in Preparation for a Pastoral Letter" (the pastoral letter itself was released by the Catholic Bishops of the Pacific Northwest and southeastern British Columbia in February 2001) uses precisely this terminology in discussing the human relationship to the natural world. "What might the Columbia River watershed look like if it were regarded and treated as a *sacramental commons, a shared ecosystem revealing God to us*?" (quoted by Robbins 2000a, 10, my emphasis).[9] "The Columbia watershed should be sacramental. It should reveal God's loving creativity in its diversity of creatures, topography and people, and its ability to provide food and shelter for its inhabitants" (quoted in *Ibid*, 11). John Hart, a member of the project steering committee, says, "A sacrament is a moment of encounter with God. To say the river is a sacramental commons means people can experience the creator in creation, outside the formal church settings" (Robbins, 2000b). This concept of the sacramental extends to human participation in the integrity of an ecosystem, which involves both appropriation and protection, and participation becomes a way of making contact with the divine. In the environmental ethics of Holmes Rolston III, hunting "is [or can be] a *sacrament* of the fundamental, mandatory seeking and taking possession of value that characterizes an ecosystem and from which no culture ever escapes" (Rolston 1988, 91). This is a secularized sense of the sacramental as a ritualized expression of and participation in essential natural processes.

Saying grace in Western religious culture is gratitude and thankfulness directed to God, the ultimate source of life. Rituals of thanks and apology in many hunting and gathering cultures are in the first instance directed to the animal killed, either before or after its death. In both cases, the gratitude is modeled on the gratitude that one person can express to another person. Does this exhaust the possibilities of genuine gratitude? How can hunting be sacramental, as some hunters describe the experience, in the absence of such spiritual dimensions? A hint can be found in Ted Kerasote's distinction between *receiving* life as opposed to merely *taking* it (Kerasote 1994, 192). For Kerasote, receiving life requires both that kind of intimacy with the country that makes one part of it rather than its colonizer, and a moment before the kill in which he apologizes to the animal. The animal's life is received as a gift, not a voluntary gift from this particular animal, but a gift from the world of which both the hunter and the animal are parts. The animal can be present as giver without the (to us) anthropomorphism of the voluntary gift. The animal does not sacrifice itself but is sacrificed, that life may continue. That life includes our life as part of a world that contains deer, elk, and other animals we hunt, as well as the many more animals we do not hunt. Gratitude—expressed both in rituals and in a practical commitment to defend the system that gives the life that is taken—is one way in which our lives can be worthy of the sacrifice.

What are we to make of Kerasote's apology in a secularized, scientifically-oriented culture? In the first place, Kerasote is not "cannibalizing" Native American culture here. His ritual is a very personal one, developed out of his own experience of having given up hunting and then returning to it. The function of the apology is not to pacify the soul of the animal, but to ensure that his own consciousness of the meaning of his actions is at its most intense. The moment of apology is a moment of both gratitude and self-discipline: it defines the *way* of hunting he demands of himself. To receive life, as opposed to taking it, requires that one have a deep sense of awe or perhaps reverence in the face of the phenomenon of life itself, embodied in the animal one is about to kill. Gratitude, which must be cultivated, and sedimented in practices and rituals, allows us to attain what Paul Shepard calls "the moment of respect and affirmation for a giving world" (Shepard 1991, 46), something that remains a possibility in the contemporary world. Such practices shape the ways we live and relate to the world of which we are a part, and ultimately this has a great deal to do with the fate of the natural world that is so threatened by human destructiveness.

Hunting as a *way*,[10] not *the* way, of giving form to our lives as a whole, even if we spend only a small part of it hunting: if hunting is to have this impact on the ways we live, on the kinds of lives we live, it has to be more than "fun," and while killing is an essential moment in the hunt, it cannot be the whole thing. Killing is the goal of the act of hunting, but it cannot be the final cause of the hunt. Only life itself can play that role, and this includes the lives of the individuals who hunt and the life of the culture in which we must determine not only whether hunting is to be tolerated, but also whether we will cultivate the kind of respect for wildlife and wild ecosystems that are necessary for any real hunting to be possible. Ultimately the meaning of hunting lies only in our affirmative relationship to life itself and to our place as mortals who live by appropriation and who ultimately die to return our bodies to the wilderness that is the earth.

The expression of gratitude or thankfulness can take many forms. Some of the traditional forms are not available to many of us, at least temporarily, in the contemporary world. But this does not mean that we cannot cultivate such attitudes, imbuing our practices with a respect that they would otherwise lack. For some this will be only the first step in a larger spiritual journey. For others, this will be their way. But a way, in order to be truly a way, must be open-ended. One can never determine in advance just where the way will lead.

Part V

The Ethics of Catch and Release Fishing

Chapter 8

Fishing for Fish versus Fishing for Pleasure: A. A. Luce and the Ethics of Catch and Release Fishing

> *ENNIS, Mont.—Jasper Thomas was flyfishing for rainbow trout on the Madison River near here one recent morning when it suddenly began raining rocks.*
>
> *Two young men were bombarding his fishing spot with baseball-size stones. "I said," 'Hey, you'll scare the fish!'" recalls Mr. Thomas, a retired Texas contractor. "They said, 'That's the point.'"*
>
> (*Wall Street Journal*, October 10, 1995)

In the summer of 1996 I fished the Yellowstone area under the careful guidance of Chip Rizzotto at High Country Outfitters. After our first day, which we spent fishing Len's Lake in Paradise Valley, and after a fine dinner of fresh salmon fillets prepared by Francine Rizzotto, we sat around the living room drinking eighteen-year-old Macallan and rehashing fish caught and fish lost. It had been a difficult day for still-water fishing, with only very light and sporadic hatches of damselflies. But each of us had landed some fat rainbows, and we were filled with the strength and beauty of the fish, the spread of the blue sky, and the striking view of Emigrant Peak we had enjoyed all day. Indeed, we had been more than happy to comply with the basic catch and release rule at the lake. That was why the salmon for our dinner had been flown into this mecca of wild and delicious trout.

As the evening progressed I found myself uneasy with my feeling of deep satisfaction and well-being. We were having a great time, but at what price? We had handled the fish carefully, releasing them unharmed—at least as best as we could tell. But what about the panic and pain our little game had caused the fish? Was that suffering a strictly negligible quantity? I started baiting my friends with my misgivings, and they rose to them as freely as cutthroats in a wilderness area, fighting just as furiously, and we spent several of our remaining evenings debating the issue. They weren't about to buy any of my sentimental claptrap, and frankly I hoped that they were right. But I kept pushing because there was a little voice in my ear that insistently asked how I could justify tormenting living, sentient beings simply for the sake of my pleasure. That voice belonged to a philosopher now dead and, to many, forgotten. His name is A. A. Luce.

Arthur Aston Luce (1882–1977) was an ordained minister as well as senior fellow and professor of moral philosophy at Trinity College, Dublin, and was best known in philosophy as the editor of the complete works of the seventeenth-century philosopher George Berkeley. But he was not just a scholar of the history of philosophy. His own thinking was imbued with Berkeley's spiritualist metaphysics, and nowhere did this come out more forcefully and beautifully than when he wrote about his avocation, fishing. Luce was an angler who could still write in the gentle spirit of Isaak Walton, and like Walton he felt compelled to raise the issue of the ethical status of fishing. His book, *Fishing and Thinking* (published in 1959 and reissued in 1993 by Ragged Mountain Press with a Foreword by Datus Proper), contains a chapter entitled "The Ethics of Angling," in which he asks if Walton is right in claiming that virtue and angling are compatible with one another.

When Walton wrote *The Compleat Angler*, he had to ask whether a virtuous man could be an angler, and he offered an impassioned and poetic defense of "the most honest, ingenious, quiet, and harmless art of angling" (Walton, 41). The question was whether angling is too frivolous an activity for a properly virtuous and God-fearing person, and it gained its urgency from the rise of Cromwell. But in 1653 it did not even occur to Walton to defend the angler against the charge of cruelty. It simply was not an issue for him. Piscator (the spokesman for fishing in *The Compleat Angler*) merely announces that he leaves the killing of otters, which he approves, to others, "for I am not of a cruel nature, I love to kill nothing but fish" (*Ibid*, 64).

In 1739, in the American colonies, the Rev. Joseph Seccombe was still compelled to offer a justification of "diversion" (along with "business"!) in general, and fishing in particular, as being innocent and "inoffensive to God" (Seccombe, 19, 15). Like Walton, he had a specific opponent in mind, in Seccombe's case the asceticism of "Popish superstition." Yet unlike Walton, Seccombe had to defend the angler against the charge of "Barbarity": "He that takes Pleasure in the Pains and dying Agonies of any lower Species of Creatures, is either a stupid sordid Soul, or a Murderer in Heart. He that delighteth to see a Brute die, would soon take as great Pleasure in the Death of a Man"(*Ibid*, 34–35). His defense is that, far from having "a murderous Tho't," the angler is simply taking what "God, the Creator and Proprietor of all, has given us to use for Food, as freely as the green Herb" (*Ibid*). The crucial point is, "He allows the eating them, therefore the mere catching them is no Barbarity" (*Ibid*). Seccombe sees that this justification of the "business" of catching fish for food does not yet justify catching them as "diversion," but he argues that this distinction does not make a difference here. "But if we consider, that the End of Business and Diversion are the same, we shall clearly conceive the Truth. The End of both are the Refreshment and Support of Man in the Service of God. If I may eat fish for Refreshment, I may as well catch them, if this recreate and refresh me. It's as lawful to delight the Eye as the Palate" (*Ibid*, 36). This piety still echoes in the opening pages of Norman Maclean's *A River Runs Through It*.

But by the early-nineteenth century, as Luce notes, Lord Byron could write in the *Don Juan*:

> *And angling too, that solitary vice,*
> *Whatever Izaak Walton sings or says:*
> *The quaint old, cruel coxcomb in his gullet*
> *Should have a hook, and a small trout to pull it*
> (*Don Juan*, Canto XIII, cvi).

In a footnote Byron adds, "No angler can be a good man."[1]

Unlike Walton, Luce thinks that we have to face the charge of cruelty head on, taking it very seriously. He approaches this question the way any good philosopher does, by attempting to draw those subtle but crucial distinctions that constitute clarity of thought, because he thinks that many accusations of cruelty are based on a failure to consider the precise meaning of the term. "[I]t is on those distinctions that the case really turns. The truth here is on a razor edge" (Luce, 171). Luce is clear about one thing: the real issue cannot be settled by a simple appeal to sentiment, and the serious objection does not come from "warm-hearted, sympathetic folk, guided more by the heart than the head" (*Ibid*, 176). Feeling alone won't do. We must think about this seriously, and thinking requires the clarity of cool-headed thought.

Luce's arguments had struck a chord with me, and although I was not prepared to agree with his conclusions, I continued to be disturbed by the feeling that I was not yet clear about things. That was why I kept bringing it up with my friends. They had strongly rejected the charge of cruelty. We had used barbless hooks and handled our fish with care. And, after all, they (along with countless others) urged, a fish is not a mammal: a fish's mouth seems to be more like a fingernail than human or mammalian flesh. This might seem plausible enough, but it also sounded a little too self-serving for comfort. When I got back home I pulled out *Fishing and Thinking* again. I needed to figure out just what constitutes cruel behavior.

Luce defines cruelty as "the voluntary infliction of unnecessary or avoidable pain" (*Ibid*, 174). Since all angling is in a suitably broad sense "voluntary" (not many people find themselves as it were angling *by accident*), we have to focus on the words "unnecessary and avoidable," and here his argument is very uncomfortable for catch and release sensibilities. "To hook trout and put them back into the water, unless they are too small to keep and quite uninjured, is to inflict pain, however small the amount, unnecessarily, and it therefore comes under the definition of cruelty" (*Ibid*, 179). So Luce concludes that catch and release fishing is cruel and therefore unethical, on a continuum with bear-baiting. But killing a fish one will eat is quite another matter.

Luce argues that whether we speak the language of religion or of biological science, we have a right and indeed a duty to kill in order to eat. He cites *Genesis* as offering a "primal permission to kill for food" (*Ibid*, 183) and notes that life itself is based on one creature consuming others, the higher eating the lower. (Eater and eaten do not, of course, form such a nice hierarchy.) This, for Luce, is the decisive point: "The primary object of justifiable fishing is to catch fish for food; there are various pleasures incidental to angling; but they cannot justify the infliction of pain or death" (*Ibid*, 180). In other words, it is not wrong to enjoy fishing, but that enjoyment itself cannot justify the fish's pain and suffering. So to fish solely for pleasure is simply wrong. After all, one might add, if *Genesis* gives permission to kill for food, it also contains the admonition to take care of the garden, to be a good steward, which at the very least means avoiding cruelty to animals.

But, we must surely ask, don't all noncommercial and nonsubsistence anglers fish for pleasure? Doesn't the very term "sport fishing" give the game away? Luce's reply is two-fold. First, he argues that the pleasure is simply irrelevant to the issue of justifying our voluntary infliction of pain: as long as we—or perhaps someone else—eat the catch, the moral principle has been satisfied. But, second, Luce thinks that this, while true, does not yet get to the core of the matter of catch and release fishing, that we have not yet made all of the essential distinctions. Thus, he writes, "our pleasure is not the proper object of our angling, and it could not per se justify the suffering and death we inflict and cause" (*Ibid*, 183), and the first clause introduces something new into his discussion.

Luce is drawing a strong distinction between what he calls the primary object of an action (what the angler aims at or strives to achieve), on the one hand, and feelings like pleasure or amusement, which may or may not accompany this pursuit, on the other hand. This allows him to argue that the very notion of "fishing for pleasure" in the sense of striving to produce feelings of pleasure as the primary object of our activity is not just wrong in the sense of being cruel, but wrongheaded. "The true angler," he writes, citing a certain Irishman named Doctor Brown, "fishes for fish, and not for pleasure" (*Ibid*, 181). Doctor Brown, whom Luce describes as "a specialist in ethics and a confirmed angler," made his point in verse:

> *I go a-fishing in due measure,*
> *An apostolic use of leisure.*
> *I grant it is my earnest wish,*
> *My keen delight, to hook a fish.*
>
> *And yet, my scholar, do not doubt,*
> *I simply go to fish for trout.*
> *For if I went to fish for pleasure,*
> *I'd miss it, and the speckled treasure.*
>
> *For pleasure is the lady coy,*
> *Teasing her faithful shepherd boy;*
> *Seek Pleasure, and she loves to flout;*
> *Forget her, and she seeks you out.*
>
> *Be taught then by the piscatorial game*
> *The difference 'twixt true and fancied aim;*

And take this moral from the art of creeling:
Just do your job, and never mind the feeling
(cf. *Ibid,* 182–183).

(For those who are used to a higher standard of poetry than this pleasant doggerel, I refer the reader to Luce's beautiful chapter on what he calls "Yeats's country" and his discussion of Yeats's poem, "The Fisherman.")

Leaving aside Doctor Brown's assumptions about men and women, coy ladies and faithful shepherd boys, this is an insight that goes back at least to Aristotle: if you are constantly seeking pleasure or happiness, it will elude you. But if you choose worthy goals and pursue them seriously and wisely, lo! you will take great pleasure in it and discover that you have lived a happy life. So one who sets out to fish for pleasure is misguided, and that particular creel will remain empty, no matter how full the wicker creel may be or how many fish are released unharmed. And indeed, our lakes and rivers are full of anglers so hell-bent on having what they think ought and must be a good time that their angling is distinctly devoid of pleasure. (I fear that not many have read Walton—or Luce—which would be a good antidote.) But will this distinction between fishing for fish and the wrongheaded endeavor of fishing for pleasure do the work Luce has assigned to it in his consideration of the charge of cruelty?

Luce is now arguing two distinct points. First, he argues that the pleasure we get from fishing cannot justify the pain and death inflicted upon fish. Only fishing for fish we intend to eat can provide that justification. But in addition, Luce is arguing that catch and release fishing, as fishing for pleasure, is misguided, since pleasure is not and cannot be our proper aim in fishing. Why do we fish for fish? With regard to the charge of cruelty, his answer is that if we are not fishing for pleasure (which is misguided anyway) we must be fishing for fish *in order to* supply ourselves or others with food. Only this will justify the suffering inflicted.

This is what had disturbed me, because it condemns catch and release angling as cruel. But as I read and reread Luce after returning from Montana, I slowly realized that at least part of his argument here won't stand up. Indeed, his entire distinction between the primary object or aim of fishing and the accompanying feelings of pleasure is irrelevant to the issue of cruelty. Luce confuses two different distinctions:

1. Fishing for fish versus fishing for pleasure—as the primary object or aim of our action.
2. Fishing for food versus fishing for pleasure—as our motive in choosing this primary object.[2]

The phrase "fishing for pleasure" has a distinctly different meaning in each of the two contexts. The first, fishing for fish versus fishing for pleasure, is a distinction related to the art of living, and instructs us as to the proper focus of our attention as we pursue our lives. Indeed, this distinction is relevant to the activity of catch and release fishing: when done properly the angler is precisely fishing for fish . . . and taking great pleasure in the entire enterprise. Catch and release fishing is by no means wrongheaded when viewed from the perspective of this distinction. The second, fishing for food versus fishing for pleasure, suggests two different explanations (which are not, of course, mutually exclusive) of *why* it is that we fish for fish. Indeed, Luce himself admits as much when he writes, "The pleasures of angling explain why people angle" (*Ibid*, 185), even if they eat their catch. This is not necessarily true of the subsistence fisherman. And anyone who simply likes to eat salmon but takes no particular pleasure in angling is best advised to buy their fillets—or cultivate the friendship of a generous angler! Luce's proper point, when it comes to the ethics of catch and release fishing, is not that it is a self-defeating mistake to focus on the pleasure we hope to get from an activity rather than on the activity itself—though this is true enough. His real point is rather that the pleasure we derive from any form of angling is not sufficient as "defence and justification," and I had to take this argument seriously.

If fishing for fish *for food* is justified, what possible justification can be offered for catch and release fishing? This pushed me back to the question that continues to engage reflective anglers: Why do we fish? Luce is right that we take an enormous pleasure in angling and that we would hardly go to the effort if this were not the case. My critique of Luce's fishing-for-fish versus fishing-for-pleasure left me with a puzzle. According to Luce, our aim in the kind of fishing he approves of is that we fish for fish in order to eat them. We do this because of the pleasure it gives us, but it is the strictly utilitarian function of nutrition that justifies our actions. But what then about catch and release fishing? What is the proper object or aim of this kind of angling, and how can it justify the suffering inflicted on the fish?

These abstract questions made me think back to those days in Montana. What made that trip such a rich experience, one that continues to resonate in my memory to this day? What was I actually after? I remember many things from that first trip: Mont's twenty-two-inch rainbow that jumped again and again at Len's Lake; the glorious thirteen-inch brown trout in the Gibbon that took my dry fly an instant after Chip muttered in my ear, "You can't make that cast any better than that!" (was it on the third or the eighth cast aimed at that little pocket?); the big trout in DePuy Spring Creek that rose to my fly after twenty minutes of casting—and the way I pulled the fly out of its open mouth as the fish turned back toward me to take it. This and more is what I remember, what I brought home with me. I experienced all of these things with great pleasure (and no small amount of anguished disappointment on occasion!), but curiously enough it is not the pleasure that I remember. I remember the fish, the lake, the clouds in the sky, the moving water, Mont getting a strike in the "Ph.D. Pool" at DePuy Spring Creek (we awarded him a Masters of Fishing degree pending his earning the Ph.D.), Paul nailing the cast that earned him the nickname "Double Haul Paul," and Jeff looking at me with a skeptical eye over his single malt while I did my damnedest to convince him that he had spent the day torturing fish. I do remember the deep pleasure I took in every bit of it, of course, but in the memory the pleasure is found at the margins of the remembered events, just as the pleasure was at the margins of the events when they happened. Luce was right about this. I wasn't fishing for pleasure, but for fish, for companionship, and for a deeper living contact with a land I was encountering for the first time and which has increasingly become a part of me, although usually I have to love it at a distance.

While I do take pleasure in catch and release fishing, to reduce this complex experience to a generic and denatured inner feeling of pleasure, which is then claimed to be an inadequate justification for my treatment of the fish, does violence rather than cast light. In *North Bank*, Robin Carey writes, "Unlike meat-fishing, which jumps firmly on a certain logic, catch and release fishing, like religion, works out into realms of faith" (Carey, 15) and it is these realms that must be explored. In short, we have to ask all over again, What is catch and release fishing? What role does it play in the lives we lead, lives that are or should aspire to be a part of the land we love? What does it mean to be responsible both to ourselves and for the bounteous land that includes, as one of its most beautiful aspects, the fish we love? What does it mean to treat those fish respectfully?

We live in a different time than did Luce, and it may well be that we have to come up with different answers to these questions than he did. But simply to pursue these questions is to live more fully. To read a serious and honest thinker like A. A. Luce is humbling in this exhilarating way, because he has the power to return us to our own lives and our own experience with new questions.

As human beings, we are a part of the natural world in many different ways, ranging from the basic necessity of assimilating energy from the world around us to practices that enable us to relate to the world in ways that feed the soul with the meaning that constitutes a life genuinely worth living. Luce leads us to raise both the issue of cruelty in catch and release angling and the broader question of the role catch and release angling can play in a life that respects both its own integrity and that of the world of which we are a part.

Chapter 9

The Practice of Catch and Release Fishing

In defense of my going afield, rod in hand, and especially with reference to those pointed questions as to why I don't simply absorb and enjoy those wonders of nature and especially riparian nature about which I am ceaselessly nattering, I have sometimes quoted Antoine de St. Exupery: "A spectacle has no meaning except it be seen through the glass of a culture, a civilization, a craft"

(McGuane 2000, 55).

A couple of years ago I had the pleasure of meeting writer and conservationist Thomas McGuane, and he told a story that made my scalp prickle. It concerned a tribe of Native Americans who live along a river in British Columbia and still rely on the annual salmon runs for a substantial part of their sustenance. The run has been declining, and it got to the point that the British Columbia game and fish authority proposed that non-native fishing be restricted to catch and release only, until the run could recover. The tribe reacted with horror at the very idea of treating the salmon in such a disrespectful manner as hooking, playing, and landing them only to let them go once the game was over. They countered that if the run was that seriously endangered, *all* fishing should be stopped, including their own.

CATCH AND RELEASE—THE HISTORICAL CONTEXT

How can catch and release fishing be a respectful way of relating to the fish most anglers claim to admire and love? This question cannot be answered without some historical context. The fishing culture described in Norman McLean's classic *A River Runs Through It* is a culture in which large limits of trout are brought home to be cooked and eaten. No one blinked an eye, and no one even thought of blinking an eye. This was the 1930s in Montana, and trout were plentiful. But late in the same decade, Lee Wulf, an innovative fly fisherman living on a trout stream in New York, wrote what has been called "certainly the most important single sentence written about angling ethics in the twentieth century" (Raines, 171): "Gamefish are too valuable to be caught only once" (Wulff, xv). The original point of catch and release fishing had to do with conservation. Raines writes, "I am convinced that without the mass acceptance of catch and release, the best bass and trout fishing in the country would have been wiped out by the next century" (*Ibid*, 178, writing, of course, about the century in which we now live). In other words, if we are going to fish for wild fish, catch and release fishing (or limits so tight that they amount to much the same thing) is imperative in at least some waters.

The anthropocentrism in this entire development, beginning with Wulff's sentence, is obvious. Wulff was not making a statement about the inherent value of the trout and what respect for that inherent value demands, but about the value and scarcity of a resource for human use. He was contrasting the value of fish as food to their value for sport (cf. Wulff, xiv). Yet Wulff already describes the spread of catch and release fishing as a kind of specifically moral progress.

> *Pride in accomplishment will always remain but, seemingly, angling is reaching a new high plane when a fisherman can spend a day on the lake or stream, catching fish and returning them to the water again, unharmed, to come home empty handed. That angler keeps no trophy to show his fellow men as proof of his prowess but contents himself with the pleasures of a day well spent in the surroundings he loves. He has fished for sport and not for glory. Upon him and those who follow his leadership the future of angling depends* (*Ibid*, xv–xvi).

What has happened in the decades since he wrote those words has in some places surpassed even Wulff's characterization of catch and release angling as "a new high plane" of angling. What began as a self-interested self-limitation has become a moral and even moralistic principle. On trout rivers where one is allowed to keep some fish, many if not most anglers, especially if they are fly fishermen, carefully release all fish. Many fishing guides have a strict catch and release policy, even if it is not legally required on the waters they fish. The result is that fly fishing magazines regularly publish letters from anglers complaining of being scorned and verbally abused by other anglers when they—quite legally—kill a few fish for dinner.

In addition, as catch and release fishing has become part of angling culture, anglers are increasingly finding themselves confronted with ethical questions that are *byproducts* of the practice, as opposed to ethical questions about the practice itself. First there was a reaction against the practice of catching large fish on tackle that is as light as possible, and then releasing the fish. The problem is that if the tackle is too light one has to play the fish to exhaustion before one can land and release it, and the fish's recovery is compromised. Now one reads again and again in fly fishing literature that one should use tackle that is adequate without being too heavy.

As fishing pressure has increased on some of the great western rivers, guides report that as the season progresses, the trout lose the strength they had in spring and early summer (when the fish are often described as being "hot")—one good run and they can be reeled in. The general assumption is that the fish are simply being caught too often, and cannot fully recover their energy. Guide and writer Larry Tullis has even suggested that such trout anticipate being released. "After hook-up, many of these fish make an initial rush with much head shaking, but then come into hand docilely with little resistance. It's almost as if they were saying, 'Come on buddy, I know the routine, let me go now so I can get back to feeding'" (Tullis, 26).

But increasingly an uglier truth has become apparent. Guides report seeing more fish with damaged and deformed mouths from being hooked and released repeatedly. I encountered this most viscerally in the summer of 2002. I was in the Lamar Valley of Yellowstone National Park for several days of wolf-watching and fishing in the famous trout streams of the area: the Lamar River, Soda Butte Creek, and Slough Creek. After setting up camp, I drove up the valley in the late afternoon, noting with some dismay the number of anglers fishing the Lamar River and Soda Butte Creek. As I walked to Soda Butte Creek in order to get closer to a group of bison, two young anglers came by, and I asked how the fishing was. Very good, one replied, but one had to keep changing flies as the fish fed on now this, now that. Without prompting he continued, "But their mouths is all tore up. They get hammered hard pretty much every day." It was immediately clear to me that I would not be fishing Soda Butte Creek or the other heavily fished waters. I was depressed for two days, and did not begin to emerge from it until I found a section of Slough Creek that was not as popular (or as full of fish) as the famous meadows. I fished it for one day, catching and releasing a very modest number of beautiful, healthy cutthroat trout (and one leaping rainbow, which I could have had for lunch if I had thought to bring my stove with me).

Even if one accepts the ethical permissibility of fishing, and of catch and release fishing, this is the unsavory side that cannot be ignored. These developments have stimulated a lot of spirited debate in fly fishing magazines and literature. Increasingly one reads the suggestion that just as in the past there were limits as to how many fish one could keep, in the future there should be self-imposed limits on the number of fish one catches and releases on a given day. My experience in Yellowstone suggests that we cannot rely on such self-imposed limits. From here it is a short step to officially mandated limitations on access to highly stressed waters. (In 2001, a bill was introduced in the Montana legislature that would authorize limiting access to certain rivers by nonresidents.)

In the early 1990s I played with the idea of cutting the hook off of my flies and turning dry fly fishing into a new practice I called "fishing for strikes" as opposed to fishing for fish. I later read that in the early 1980s a fisherman named Mike Kimball began to use hookless flies as a research tool: a fish that takes such a fly and spits it out continues to feed, enabling Kimball to switch flies and test new patterns. Kimball's friend Jim Emery then generalized the practice for ethical reasons: under heavy fishing pressure, catch and release fishing "amounts to serial harassment" (Bodo, 34). A problem with this approach is that it is really only applicable to dry fly fishing. One fly fisherman has addressed this by producing hooks that have second eye in place of the pointed end of the hook. Unlike the hookless fly, this allows the angler to make contact with the fish for at least a few electric moments, though the fish can easily throw the fly if allowed a bit of slack in the line. This approach can be extended to nymph and streamer fishing.

These are ethical issues and solutions that arise *within* catch and release fishing, problems produced by the practice itself. They are grouped around the problem of stress and damage to fish, not pain, as in A. A. Luce's ethical reflections. But is the shift of emphasis from pain to stress and damage more than an attempt to avoid the real moral issue in catch and release fishing? If catch and release fishing is a condition for having wild fish to fish for at all, why should not the ethical alternative be to stop fishing completely? Stephen Bodio writes, "Hunting is a play about life and death and the transfer of energy, but if we stray too far from eating, I question the seriousness of our play" (Bodio 1998, 85). By definition, catch and release fishing cuts the tie between the practice and the structure of appropriation that is both basic to our being and, I have argued, fundamental to the moral status of hunting. If our play lacks seriousness, then, it would seem, we are merely playing and should grow up.

At the very least, I think that these considerations point to the fact that catch and release angling becomes morally questionable when it is taken to be the expression of an ideal of personal moral purity, rather than a necessity imposed by the scarcity of a resource. Fishing, all fishing, including careful catch and release fishing, is a blood sport, with all of the seriousness and responsibility that involves. I have released a trout in a lake where fishing was restricted to catch and release, only to see blood from the fish's gills staining the water. I knew that the fish was mortally injured—just how it happened I do not know—and it was dead in less than a minute. The rules did not allow me to kill fish intentionally, but the fish was dead. Unfortunately, I was in no position to cook the fish for supper, so I retrieved the carcass and threw it up in the brush that bordered one part of the lake, knowing that it would provide food for a scavenger. My responsibility for the fish's death cast a pall over an otherwise glorious day, a pall that would not have been cast if I had deliberately killed the fish for dinner. Catch and release is a discipline, an imperfect discipline that we impose on ourselves out of necessity. It is not a way to fish while keeping one's hands clean, if that is what one desires. Rather than disburdening us of responsibility, it increases the load. As Bryn Hammond writes, catch and release should be practiced "only on waters where it makes real sense and not simply imagined sense. Where and when it is practiced, it should always be for the sake of the trout, not for the sake of the angler. If any benefits accrue from its employment then they are due to the fish alone. If not, then it becomes an act of selfishness" (Hammond, 121). This presupposes the moral permissibility of catch and release fishing, but it means that when the fish begin to suffer permanent harm as a result of this practice, we have to fish less, or not at all. It also means that some waters managed on a strict catch and release basis, especially private waters, might require a change in their regulations. Managing a lake for catch and release fishing for trophy-sized trout may well be ecologically imbalanced, something that undermines the integrity of the fishing.

But the question of the general moral permissibility of catch and release fishing remains, and some avid anglers have come to the conclusion that pure catch and release fishing is inconsistent with a genuine respect for the fish. John Holt writes,

> *I'm with the Inuit and the Dene of the Yukon and Northwest Territories on how they feel about this issue. To catch fish and not kill and eat* any *of them is both disrespectful to the fish and a betrayal of the angler's heart. Don't get me wrong, though. Catch and release* does *help maintain a fishery in some overworked waters, and in these situations it has its place. But you won't find me on those waters anymore* (Holt, 99–100, my emphasis; cf. also McCully, chapter 5).

I turn now to a direct discussion of the two issues Luce left us with: the issue of cruelty and the issue of the relationship of catch and release fishing to an attitude and practice of respect for nature as developed in Part IV above.

CRUELTY REVISITED—THE SCIENTIFIC PERSPECTIVE

> *Imagine using worms and flies to catch mountain bluebirds or pine grosbeaks, or maybe eagles and ospreys, and hauling them around on fifty feet of line while they tried to get away. Then when you landed them, you'd release them. No one would tolerate that sort of thing with birds. But we will for fish because they're underwater and out of sight.*

The writer is not a representative of People for the Ethical Treatment of Animals, though PETA representatives are fond of similar analogies.[1] The writer is Jack Turner, philosopher, mountain guide, eloquent writer, and "former angler" (quoted in Kerasote 1997, 24). In the past several years, a number of voices have been raised, following in the footsteps of A. A. Luce, questioning the morality of catch and release fishing. The arguments I am interested in here are not necessarily made from a strict animal rights or animal liberation point of view, since many of those making the arguments have no objection to catching and eating fish. The argument is rather that conservation and wildlife management is not an adequate defense if the practice is cruel in the sense of inflicting unnecessary pain. Rather than catch and release, some are calling for a "catch-and-kill ethic" in which one fishes only for dinner, and stops fishing once one has caught enough (Talbot, 16).

Jack Turner is perhaps the most profound philosopher of wildness currently thinking in the tradition of Thoreau, and when he makes common cause with PETA, I for one have to take him seriously, since I know that he is not just looking for an inviting piece of rhetoric to represent a pregiven, dogmatic position. Just as Ted Kerasote thought his way back to a certain kind of hunting after years as a vegetarian, Turner seems to have thought his way to the requirement that he cease participating in a practice he deeply appreciated. More directly, if the analogies drawn between fish on the one hand, and birds, kittens, and human beings on the other hand, are accurate, if it is true that catching and releasing a fish inflicts anything like the pain and suffering that would be caused in the analogous activities, then not only does catch and release fishing stand convicted of cruelty, catching fish for dinner using a hook and line may be immoral as well. I think that most anglers do think that there is an important difference between mammals and fish, and that the difference makes a difference in what is morally permissible in our practices. The question is, just what is the difference, and why does it make such a difference?

In his exploration of the charge that catch and release angling is cruel, Ted Kerasote turns to the scientific evidence offered by Michael K. Stoskopf of North Carolina State University that fish feel pain. In a journal article published in 1994, Stoskopf offers two primary lines of evidence, one behavioral and the other biochemical, for the claim that fish feel pain. First, Stoskopf argues that fish exhibit the same responses to noxious stimuli as do mammals: "elementary rapid startle reactions, simple nonspecific flight, and affective responses such as vocalization" (Stoskopf, 775). A fourth basic response is "coordinated reaction, presumed to be controlled in the cortex, such as biting the source of pain or rubbing the site of stimulation" (*Ibid*). The general principle used for interpreting these responses is that "the variability of response to noxious stimuli can be considered a marker of sentience" (*Ibid*)[2] Stoskopf also refers to "the use of learned avoidance behaviors to assess pain" (*Ibid*). His general conclusion, directed to scientists doing research on fish, is that "current knowledge suggests that it is appropriate and preferable to consider procedures known to be painful when applied to human subjects to be painful when applied to other animals" (*Ibid*). He particularly emphasizes that "it would be an unjustified error to assume that fish do not perceive pain . . . merely because their responses do not match those traditionally seen in mammals subjected to chronic pain" (*Ibid*, 776).

Second, Stoskopf turns to the biochemical evidence. "The biochemical evidence for pain perception in fishes is compelling. Teleost (bony fishes) nervous systems produce compounds related to andrenocorticotropic hormone and opiates" (*Ibid*).[3] Stoskopf argues that the presence of opiates and receptors for other substances involved in the "biochemical mediation of pain" in mammals is evidence for a similar function in fishes (*Ibid*). In short, continuity in behavioral responses to noxious stimuli, along with the presence of the same opiates that reduce pain in mammals, is evidence for the claim that fish feel pain.

Stoskopf's evidence is an example of, or better perhaps, it rests upon, what Charles Darwin called "evolutionary continuity." "For Darwin . . . evolutionary relatedness signifies an unbroken continuum between humans and animals in all respects, including behavioral pattern *and mental* faculties" (Crist, 18; my emphasis). As Darwin saw it, continuity in behavioral patterns is evidence of continuity in *mental faculties*, leading to the conclusion that "there is no fundamental difference between man and the higher mammals in their mental faculties" (Darwin, 1:35; cf. Crisp, 309–320; Allen and Beckoff, 22–25). While Stoskopf is not arguing that there are no fundamental differences between fish and higher mammals, he is pointing to similarities that seem to be evidence of a shared ability to feel pain. Thus, the behavioral and biochemical evidence Stoskopf calls on is just what we would expect from an evolutionary point of view. Together they seem to make a compelling case for the claim that fish feel pain. Stoskopf himself does not use this evidence to make an argument against fishing; he simply thinks that it is important that we all be clear about what is really going on. But it does seem that Stoskopf's evidence might well be sufficient to convict the catch and release angler of cruelty.

Yet things may not be as simple and straightforward as they seem here. In an essay entitled "Do Fish Feel Pain?," published in the fishing magazine *In-Fisherman*, James D. Rose of the University of Wyoming argues that we have to be careful to make a distinction between behavioral and biochemical reactions to injurious stimuli on the one hand, and pain as a psychological experience on the other (Rose 1999, 39). The significance of this distinction, as I see it, is that evolutionary continuity does not justify a *direct* inference from behavioral and biochemical reactions to injury on the one hand, to the presence of the psychological experience of pain on the other hand. Once this distinction is drawn in principle, Stoskopf's argument commits the fallacy of a *simple* inference from biochemically-induced behavior to psychological experience. As Rose argues, "in reality, reactions to nociceptive [injurious] stimuli are protective responses that can occur in forms of life that are incapable of perceiving pain" (*Ibid*). A generally accepted example of such forms of life would be insects.

This would seem to lead to a stand-off, with each side arguing that the other bears a burden of proof that has not been provided. As I read them, both Rose and Stoskopf accept the distinction between behavior and psychological experience. Stoskopf can appeal to evolutionary continuity: the same biochemical compounds that inhibit the experience of pain in mammals are found in fish, and there is a presumption in favor of similarity of function, thus shifting the burden of proof to Rose. Rose can reply that evolutionary continuity does not guarantee a correlation between an organism's behavior (or biochemistry) and that organism's psychological experience. Indeed, Rose has emphasized that the shared evolutionary history between mammals and fish ended about four hundred million years ago (Rose, personal communication, 3.28.2001). How, then, can one show that the *distinction* between behavior and biochemistry on the one hand, and psychological experience on the other, leads to a justification of Rose's claim that some forms of life, including fish, cannot feel pain? How can one show that some forms of behavior *are* expressions of psychological experiences such as pain, while similar behaviors in other species is not associated with psychological experience?

Rose builds up his argument for the claim that fish do not feel pain slowly. He starts by making three points about physiology:

1. In human beings, some of the reactions to nociceptive stimuli are generated at the spinal cord level. Other reactions are generated by the brainstem.

2. In human beings conscious awareness of nociceptive stimuli depends on certain parts of the cerebral hemispheres. If the neural activity does not reach the cerebral hemispheres, there is no conscious experience of pain [though there could still be a behavioral reaction generated by the brainstem].
3. "The fish central nervous system . . . consists of a simpler version of spinal cord and brainstem [compared to humans], with only small and primitive cerebral hemispheres. The neural functions that generate behavioral responses to a nociceptive stimulus are much the same at the spinal cord and brainstem levels as in a human. But the cerebral hemispheres of fish lack the regions necessary for conscious awareness and for generation of a pain experience. *So* no awareness of pain is associated with the brainstem and spinally generated behavioral reactions" (Rose 1999, 39 box; my emphasis).

In short, fish cannot feel pain because they do not have the structures in the cerebral hemispheres required for the experience of pain.

Now a skeptic might answer that the crucial "So . . . " is too strong. Just because humans don't experience pain unless certain regions of the human cerebral hemispheres are stimulated does not prove that the experience of pain in fish has to be dependent on strictly analogous structures. So the absence of those structures proves nothing, especially in light of the fact that fish, unlike insects and earthworms, do indeed have cerebral hemispheres. Therefore, given the behavioral and biochemical evidence Stoskopf describes (the presence of opiates, etc.), it would seem reasonable to infer that fish do indeed experience pain.

Rose has a powerful counterargument at this point.

> *If the cerebral hemispheres of a fish are destroyed, the fish's behavior is quite normal in most ways, because the simple behaviors of which a fish is capable*—including all of its reactions to nociceptive stimuli—*depend mainly on the brainstem and spinal cord. A human's existence is dominated by the cerebral hemispheres, while a fish is a brainstem-dominated organism. Therein lies the answer to whether fish feel pain* (*Ibid*, 41; my emphasis).

Note that both the story Rose is telling as he denies that fish feel pain and the story Stoskopf tells in arguing that fish do feel pain are perfectly consistent with evolutionary continuity[4] (though only Stoskopf's argument needs to appeal to the principle as justification of his position). More sophisticated neurophysiological structures evolve on the basis of, and continue to depend on, less sophisticated structures. The cerebral hemispheres depend on the brainstem, which is a more primitive neurophysiological structure.

There is of course a puzzling aspect to the fact that a fish's behavior is largely unchanged if its cerebral hemispheres are destroyed: what, then, are the functions of the cerebral hemispheres in fish? Do they have no function? That would be highly implausible. Rose does not answer this question in the essay I have been discussing, but he does in a much longer and more technical version of the argument (Rose 2002). When their cerebral hemispheres are destroyed, fish lose their sense of smell and some of their ability to modulate behavior. The latter makes avoidance learning much more difficult, though most forms of learning remain intact (Rose 2002, 9). Complete removal of the cerebral hemispheres in fish leaves behavioral responses to nociceptive stimuli—to injury—completely unchanged.

There is another aspect of Rose's argument—this time concerning biochemistry—that may seem puzzling, especially from the point of view of evolutionary continuity. As noted above, Stoskopf points out that the nervous systems of bony fishes produce compounds related to those that mammals produce when subjected to pain. These compounds include endorphins, which are opiates; and why would a nervous system secrete opiates if not to control pain? Philosopher David DeGrazia has made this argument explicit, though his language is careful:

> *In the brains of vertebrates, the neural mechanisms implicated in what is apparently pain behavior are very similar across vertebrate species (including humans). Anesthesia and analgesia control what is apparently pain in all vertebrates and some invertebrates; in all vertebrates, the biological feedback mechanisms for controlling what seems to be pain—involving serotonin, endorphins, and what is known as* substance *P—are remarkably similar* (DeGrazia, 109).

But DeGrazia also notes that one cannot in any simple way infer the ability to feel pain from the presence of such naturally-occurring opiates. Such substances are found in insects and earthworms, and DeGrazia himself comes to the conclusion that they are incapable of feeling pain. "It does not immediately follow, of course, that insects and earthworms feel pain, *only that their nociceptive pathways might be inhibited by opiates*" (DeGrazia, 110; my emphasis). This is important, since all that is required from the point of view of the theory of evolution is that such opiates—which in humans function much the way Novocaine does, by inhibiting the nociceptive pathways that in humans lead from the tooth the dentist drills to the neocortex (Rose 1999, 39)—can have survival value by affecting the organism's reaction to injury. The logic of the principle of evolutionary continuity does not require that just because an organism secretes opiates it necessarily feels pain. Rose argues that in fish just the opposite is the case, since in fish such biochemical inhibition occurs at the level of the spinal cord and brain stem, and not at the level of the cerebral hemispheres, much less that of a neocortex.

Indeed, the idea of evolutionary continuity might suggest just the opposite of Stoskopf's interpretation: in order for the awareness of pain to develop evolutionarily, a *lot* of neurophysiology has to be already in place—physiology that will play various important roles in the awareness of pain, when it develops. While it is appealing to think that first awareness of pain developed and then certain mechanisms that have the function of controlling that pain developed, this turns out to be mistaken. It is therefore evolutionarily anachronistic to suggest that the presence of these latter structures is evidence for the presence of pain awareness. Structures can acquire new functions in addition to their original functions. I am not arguing that the issue can be settled *a priori*, but only insisting that all of the relevant facts be taken into account.

Rose's basic criticism of the work of scientists such as Donald Griffin (and by implication that of Stoskopf), who have argued for the existence of conscious awareness in many animals, is that they do not give adequate consideration to the question of feasibility, where "feasibility" refers to the neurophysiological structures that underlie and make possible all behavior and all psychological experience.[7] Rose invokes "one of the most well established principles of neuroscience: that functions of nervous systems, including psychological functions, depend on specific structural features of these nervous systems" (Rose 2002, 29). If it can be shown that an organism does not have any neurophysiological structures capable of making a type of psychological experience possible (as opposed to their simply not yet having been discovered), arguments that the organism actually has such experiences are *prima facie* implausible.

In other words, Rose fills in a bit of the gap Stoskopf points to when he writes, "The question of central processing of pain by fish remains open" (Stoskopf, 776). Since the fish's reactions to nociceptive [injury] stimuli are unchanged by the removal of the cerebral hemispheres, central processing in the hemispheres cannot control that behavior. Similarly, Stoskopf's use of coordinated reaction to noxious stimuli as evidence for the claim that fish feel pain fails, since he presupposes that such reactions are "controlled in the cortex" (*Ibid*, 775), which turns out not to be the case. Stoskopf could still argue for the general "use of learned avoidance behaviors to assess pain" (*Ibid*), since it is precisely the learning of avoidance behavior that is affected by removal of the cerebral hemispheres. But the fact that what Stoskopf interprets as pain behavior occurs independently of functioning cerebral hemispheres, and other forms of learning are unimpaired, seems to raise more questions than it answers. More importantly, not all avoidance learning presupposes conscious processes (Rose 2002, 8–9). Rose's point here would be that Stoskopf's use of learned avoidance behaviors to assess pain begs the question.

Dawn Carr, Fishing Campaign coordinator for the People for the Ethical Treatment of Animals, responded to an Associated Press article on Rose's work by writing a letter of protest to Philip L. Dubois, President of the University of Wyoming, demanding that Dubois "dismiss Rose's claims as nothing more than a fish story—and retract his studies before they damage the institution's reputation" (Carr). Carr appeals to the statements of several scientists, including Michael Stoskopf. In his reply, President Dubois, noting that he is a political scientist (and fisherman), not a neuroscientist, was careful to affirm his institution's commitment to academic freedom.

President Dubois was certainly correct in refusing to take an institutional position concerning an issue of scientific research, and Carr's demand should be taken for what it is, a piece of political rhetoric (and perhaps an effective one). Further, as a political scientist, Dubois was correct not to assume that he is in a position to make a considered scientific judgment about technical issues in a quite different field. But an angler ought to be concerned with the issues under debate here. I am not a neuroscientist, but as a philosopher and an angler, I have tried to sort out what the issues are when we bring the findings of neuroscience to bear on the question of fish pain, and to consider the evidence, which can and has been presented in a manner accessible to the layperson. The result has been that for the moment, at least, the angler's prescientific intuition that we must beware of false anthropomorphizing when it comes to fish has been confirmed: there is indeed a difference between fish and mammals here, and there is reason to think that it does make a difference. The question I want to ask now is whether this is or can be the end of the story.

FISHING THE LIFEWORLD

In a recent book, Eileen Crist studies the ways in which the language we use has an effect not only on how we understand animal behavior, but also on how we perceive animal behavior (Crist, 206). Her book focuses on the ways in which, as she puts it, the language used has an effect on the "different portraitures of animals" (*Ibid*, 1) in the work of Charles Darwin, of turn-of-the-century naturalists, and in works of classical ethology and contemporary sociobiology. Countering the prejudice that the use of anthropomorphic language is a "mere metaphor or category mistake" (*Ibid*, 203), she first traces out the way in which Darwin used his principle of evolutionary continuity to justify anthropomorphic language, which is very close to everyday language and common sense. She then goes on to show how such language offers a coherent mode of access to a world rich in subtlety and meaning (*Ibid*, ch. 1). Phenomenologists have called such a world a "lifeworld," which Crist characterizes as "an everyday world where things, activities, relations, and events have experiential significance" (*Ibid*, 54). Darwin's assumption, based on the continuity principle, is that animals inhabit lifeworlds. Other conceptual frameworks, such as that of ethology, use a radically different language, modeled on the physical sciences rather than drawing on anthropomorphic language, in constructing a coherent approach to a very different world, even as they study the very same behavior Darwin described. Crist's thesis is that none of these frameworks offers a "neutral language," since each produces its own "perspectival effects" (*Ibid*, 209). Crist's work on linguistic frameworks as perspectives is relevant to the issues I am concerned with because it raises the question as to what framework we live in as anglers. While more technically scientific perspectives—such as that of neuroscience—are necessary in considering the issue of pain, for the reflective angler the investigation of such perspectives are journeys of exploration into foreign territory. But we always return to the world of our prescientific experience, our lifeworld, which is where we live our everyday lives. We may return with discoveries that help us to better understand our practices and ourselves, but their ultimate meaning for us has to do with the way we integrate them into our own prescientific experience and the language of that experience. Even if we accept the scientific evidence as described above, it may be that, as persons, as opposed to scientists, the Native American tribe in British Columbia knows something that is beyond, or better, other than, that scientific truth, something that is essential to wisdom concerning the art of living. As human beings we live in the lifeworld, and here we relate to other human beings not as neurophysiological systems but as members of a shared world, and we relate—or at least we can relate, since not all do so—to animals as beings whose actions have a significance to the animals themselves and to us as persons. As Eileen Crist puts it,

> *The language of the lifeworld—with its emphasis on the ceaseless cascade of actions and a world where meanings are shared—is, prototypically, the everyday language of human affairs. In transferring this language to animal life, its qualities as they pertain to the human context are assembled in the case of animals as well* (Crist, 202).

We live in the lifeworld, but in the lifeworld we experience non-human animals, including fish, as living in their own lifeworld or surrounding world (*Umwelt*). Barry Lopez is speaking of animals as we can encounter them in the human lifeworld when he writes, "... our relationships with wild animals were once contractual—principled agreements, established and maintained in a spirit of reciprocity and mythic in their pervasiveness. . . . these agreements derived from a sense of mutual obligation and courtesy" (Lopez 1991, 381; cf. Part I). It is only in a lifeworld that we can experience "the awe and mystery that animals excite" (*Ibid*, 384). This is important because it helps us understand the question when we ask why the activity of fishing is important enough that it can justify catch and release fishing. Fishing takes place in the lifeworld, and must be analyzed and justified in terms of the lifeworld. The scientific evidence will only take us so far.

Steven J. Meyers arrives at a similar conclusion.

> *Anthropomorphism is the biologist's enemy, it is the scientist's nemesis; yet, it is one of the humanist's greatest strengths. . . . Poetry does not deal in verisimilitude. Neither does life. It deals in metaphor. It does not always deal in literal truth. It often deals in suspicions. All thinking humans are poets; hunting and fishing are poetic acts. Whether the apparent panic we see in game, the apparent fear, the apparent pain are, in fact, panic, fear and pain in the human sense or not (and it is doubtful that they are), our sensitivity to something akin to human panic, fear, and pain in our prey forces us to take seriously the very serious act of catching or killing game. It forces us to struggle with the morality of our behavior. In the absence of that struggle, hunting and fishing lose a necessary and appropriate element of gravity. Hunting and fishing conducted in the absence of this moral complexity demean both predator and prey* (Meyers).

So the question remains, why fish?

The "Why do we fish?" question arises even more often in fly fishing literature than the "Why do we hunt?" question in hunting literature. Better anglers, writers, and thinkers than I have had their tilt at this windmill. They have come up with many wonderful, beautiful, and true answers, and I am not arrogant enough to think that I can better their efforts. But I cannot avoid the question either. Part of the answer has to do with the nature of the fish, and with the way we experience both ourselves and the fish in this practice. Even after a careful reading of Rose's work, and of other works on fish physiology, when I cast to a wild cutthroat trout, I am not casting to it *as* a spinal-cord-and-brain-stem. When I stalk, cast to, play, and release (or kill) a fish, I am relating to what I experience as a being that is highly sensitive to its surrounding world. In presenting an artificial fly to a trout, I am entering what I experience as the surrounding world of the fish in a way only fishing allows. If I rigorously schooled myself to see in trout behavior nothing but a series of automatic responses, something would be lacking in my experience, and I suspect that I would soon cease to fish. The world would have become a less interesting place in which to live.[6] We need other concepts to understand what angling is all about.

In *A Sand County Almanac*, Aldo Leopold writes of the "imponderable essence," the "motive power," of a particular landscape, calling it the "*numenon* of material things" in contrast to the "*phenomenon*, which is ponderable and predictable"[7] (Leopold, 137–138; cf. Callicott 1989c).

> *Everyone knows, for example, that the autumn landscape in the north woods is the land, plus a red maple, plus a ruffed grouse. In terms of conventional physics, the grouse represents only a millionth of either the mass or the energy of an acre. Yet subtract the grouse and the whole thing is dead. An enormous amount of some kind of motive power has been lost* (Leopold, 137).

Leopold mentions some other examples, and while trout are not among them, I doubt that he would have disputed their rightful inclusion. Such *numena* are quintessentially inhabitants of the lifeworld, and of specific cultural-historical lifeworlds. But then, surely trout can be the *numenon* without our having to fish for them.

Some years ago David Quammen wrote a beautiful essay on trout as the *numenon* (not his word) of Montana. Why, asked his East Coast friends, do you live in Montana? "'The trout,' you answer, and they gape back blankly." Then a flash of recognition: Oh! You live there so you can go fishing, they say. Maybe at the beginning, he replies, but not so much now. So why do you stay, for something else? "'Yes. The trout,' you say. . . . 'The trout is a synecdoche,' you say . . . " (Quammen, 19–20). The point is that the trout represents itself, but more than itself. The trout stands for the whole of which it is both a part and a privileged expression. As too many fishermen have noted to quote any one of them (Quammen's essay is a good example), the places where trout occur naturally are typically beautiful places, with clear, cold, clean water. In such landscapes there are many candidates for the role of the synecdoche, the *numenon*, but for some of us there is no contest, there are no rivals.

For some reason, the trout bug bit me early on, long before I caught one. I still have a very vague recollection of our parsonage kitchen in Walhalla, South Carolina, near the foot of the southern Appalachian Mountains. My father had gone trout fishing with some members of the congregation of the Methodist church where he was pastor—probably to the East Fork of the Chattooga River, where he caught the largest trout of his life, a sixteen inch rainbow, or maybe to the Whitewater, where my grandfather fished in earlier years, and where I would enjoy a fishless day with my father many years later—and had brought home fish that were being cleaned in the kitchen. I might have been four or five years old, and it was clear to me that this was a special occasion. In the southern Appalachians, trout are generally referred to as "mountain trout."[8] These were clearly special fish, since I already knew that mountains are special places. Some years later, on a church outing at a South Carolina state park near Columbia, far from the mountains, I used all of the film in my new Brownie camera trying to take pictures of fish in a small, warm-water stream. I was convinced that they were trout. That was impossible, and all the developed pictures showed was the glistening surface of a nondescript pool, but I remained fascinated by the fish, the pool, and even the photographs. Behind the glare there was magic.

When I was about twelve my mother took us kids camping on the east side of Mount Pisgah in North Carolina, getting up early to take me to the check station where one registered to go trout fishing in the stocked waters of the Davidson River. I had no experience in fishing for anything other than small bream and carp, and after some hours of exhilarated frustration, an older fisherman took pity on me and gave me a rig consisting of a willow-leaf spinner with a fly attached. Slowly pulling it up through a riffle, I repeatedly got the tap-tap-tap of a fish, and finally hooked my first trout. I had no landing net, so I began to maneuver it over to the side of the stream, only to lose it. The moment was charged with electric disappointment. It was to be the last trout I saw for some years, not because I didn't fish, but because I just was not very good at it.

Any water, especially moving water, with fish, catchable fish—especially trout—draws me like a magnet. But the magic can appear in other waters as well. I love snorkeling in tropical waters, watching the brightly-colored fish. But for all their beauty, in my experience they have no greater claim to be the *numenon* than does the coral, and watching them does not grab and focus me the way catching sight of a bonefish feeding its way across a tidal flat does. Bonefish are the *numenon* of tidal flats because I can stand in a boat all day, straining to see the almost invisible fish feeding toward me, casting to them, and, if I am lucky enough, standing in breathless awe as a hooked bonefish races across the flats. Trout (and salmon when they are present) are the *numenon* of cold-water streams and rivers, and of the mountains that often surround them, in part because I can fish for them in a beautiful way. This is imponderable because there is no adequate answer to the question, Why this? There is no way to say why it is *this*, just *this*, that captures this particular person's imagination and passion, while for another person it remains just . . . a cold fish. And while I would love to drift down a river on Vancouver Island with snorkel gear, well insulated by a neoprene wet suit, watching the salmon as they move up the river to spawn,[9] that would not replace the moment of connection I someday hope to experience when a wild salmon takes my fly. It is the possibility of the latter experience that would be the source of the "motive power" of the former experience.

Ultimately the only answer is to be found in the experience itself, the kind of experience Henry David Thoreau had on the top of Mount Kataadn, where he encountered a wildness so other, so alien, that he was put in contact with both himself and the world with a new intensity. "What is this Titan that has possession of me? Talk of mysteries! Think of our life in nature—daily to be shown matter, to come in contact with it—rocks, trees, wind on our cheeks! The *solid* earth! The *actual* world! The *common sense! Contact! Contact! Who* are we? *Where* are we?" (Thoreau, 525). The experience of contact that occurs when a trout takes my fly is powerful enough to have started me on the road that has led to the writing of this book (the chapter on A. A. Luce was the first thing I wrote, before I had any idea of writing the rest).

The sense of "transcendent congruency" with the land, of "a state of high harmony or reverberation," that Barry Lopez evokes (Lopez 1987, 297), can come in many ways. Too often today it comes only from individual practices, practices we pursue now and again, and not from our lives as a whole. But those practices are important, to us and to the world.[10]

PRESENCE AND RELEASE

It is at this point, I think, that the difference between fishing and hunting in the everyday sense of the word becomes visible. In Part IV I mentioned some attempts to conceive of a hunting analogue of catch and release fishing, and noted that the result seemed less a new form of the practice of hunting than something else, something different, a disrespectful toying with wild animals. Why does catch and release fishing seem—to some—to be different from what Robert F. Jones calls "zap-and-release" hunting? I think that there is something very real here, that there is a difference between what we generally call "hunting" and fishing.

In most forms of hunting, the prey is not itself a predator. When the hunter's prey is itself a predator, the hunting is almost always pure trophy hunting of the form that Stephen Kellert calls "dominionistic."[11] What makes catching and releasing a cutthroat trout different from shooting an elk with a paint-ball (or a laser gun)? I think that the difference is that angling is in an important way different from almost any other form of hunting in that the fish, in order to be caught, must act as predator. (I do not regard snagging, which is a legal method of taking some kinds of fish in Missouri, to be a form of angling.) To fish is to put something out, be it worm or dry fly, in such a way that a fish will be tempted to eat it. The fly fisherman stages the role of the prey only to turn the tables, becoming predator to the fish. The catch and release angler, especially the fly fisherman, enters the world of the eater and the eaten in a highly ritualized manner. I think that this, along with the fact that fish put up a sometimes magnificent resistance to being captured, is why one can release the fish without the sense of lack of completion, the sense of having trivialized something serious.

To some, the release of a fish can have a positive meaning that is very different from seeing a splotch of paint on a departing elk. The very sense of empathy that led Guy de la Valdéne to stop hunting the woodcock he loves (see Part IV) is part of the appeal of catch and release fishing to many anglers. "I love the feeling I get when the fish realize they're free. There seems to be an amazed pause. Then they shoot out of your hand as though you could easily change your mind" (McGuane 1999, 103). This is the language of the lifeworld, the language of unapologetic and appropriate anthropomorphism. Nothing in the science of neurophysiology can destroy this experience. And, of course, there *is* a moment when you could change your mind. So is this the expression of an anthropocentrism, a royal dominion that is so sure of its own power that it can disdain to exercise it? Or is it a regenerating and respectful dip into the stream of wild life of which we are all, in the final analysis, a part? Part of the answer will depend on whether such a practice enables a genuine encounter with both the fish along with its world, and with ourselves as part of a shared world. The question is whether the catching and releasing of a fish can be an experience saturated with respectful participation, with reality.

Not everyone will agree that this is the case, of course. Sydney Lea writes,

> *Conservation instincts, or plain common sense, dictate that we release the trout, the salmon, and even the bass we now catch. Thus the gesture of angling is ever more symbolic, less what I call a primary act than somehow a charade of the primary. I might say that hunting is to contemporary sportfishing roughly as the dead black duck at my feet is to the photo of him I might have taken instead. A direct, and undiluted* presence *on the one hand, and on the other a kind of disembodied* representation (Lea, 26).[12]

Lea may have a point here, but I think it is a very limited one. The trout whose presence and vitality are telegraphed so electrically up my line and through my rod to my hand, the trout in my net, the trout whose very life is (carefully!) in my hand (many fish can be released without touching them), the trout whose color and form I admire so briefly before releasing it, is anything but a representation. Eating is not the only mode of Thoreauvian contact, as any good bird watcher can attest. In releasing a beautiful fish there is a completion in comparison to which paint-ball or laser-gun "hunting" is a bloodless husk of an experience.

Can fishing be *a way* or an intrinsic part of *a way*—the way of the trout? For Ted Kerasote, hunting is an integral component of the integrity of his way of living as a participating part of his ecosystem. Can catch and release fishing support this burden of meaning? There is no doubt that for many anglers, fishing, and especially fly fishing, is a way, something that indelibly gives shape and form to one's sense of self and world.[13] The opening words of *A River Runs Through It*, "In our family, there was no clear line between religion and fly fishing," are only the most famous expression of this (hard to read, today, without feeling a sense of cliché, at least until one is swept up again by the power of Norman Maclean's writing). John Gierach's evocation of the "trout bum" is another, perhaps less exalted, one. And catch and release fishing can be a part of a practice that does give shape to our lives and to our relationship to the natural world.

But to call catch and release fishing as such a way is to make the catch and release angler morally superior in a way that distorts the practice. Catch and release is morally required of anglers in certain situations in which we increasingly find ourselves. In those situations, it is the right thing to do, but that does not make the catch and release angler morally superior to the angler who kills some fish to eat, and does so in a way that respects and participates in the integrity of the system of which both fish and angler are a part.

And yet, even if catch and release fishing cannot as such be a way, if it is to be defended against the charge of cruelty it has to be more than conservation. Regardless of the scientific facts about pain, to catch a fish is to subject it to stress, whether one views fish from the perspective of the lifeworld or from the perspective of science. And, after all, there is always the option that we stop fishing in certain waters. Many of the small mountain streams in the Smoky Mountain National Park, home to the last native brook trout in the region, are permanently closed to fishing, and should be. But beyond such ecologically-based necessities, catch and release is more than tormenting fish for our amusement. It is a way of making ourselves a part of the land, of participating in the cycles of the land—the changing hatches of insects, the changes of the seasons, the changing moods of trout, water, and sky—even if some of the participation, in the loss of the eating of the catch, has become symbolic. Does that make the fish a disembodied representation, as Lea charges?

In a recent essay, "Fish Stories," Thomas Lynch suggests the way in which catch and release can resonate with the deep currents and structures of our lives. Recounting his son's first fish, which the boy wanted both to show to his mother, which would mean that the fish would die (and be eaten), and to release unharmed, Lynch writes,

> *He wanted to keep it. And he didn't want it to die. And I could see in his bright blue eyes the recognition that these aims were at cross-purposes. This was a game he couldn't play for "keeps." He was crying when he put it back in the water. Catch and release, like love and grief, are difficult notions. We've been fishing together ever since* (Lynch, 152).

The rhythm that dominates Lynch's essay is not that of fish-catch-kill-eat, with the incorporation, organic renewal, and reciprocity that is its issue, but rather "love and grief, sex and death, the heart's divisions between catch and release" (*Ibid*, 155). "Time seems a constant game of catch and release. Dream and vision, memory and reflection—each is an effort to hold life still, like paintings of fruit and flowers on a table. There is no still life" (*Ibid*). Here the "game of catch and release" takes on the cosmic—and existential—dimensions I found earlier in Stephen Bodio's image of hunting and fishing as "elaborate ways of playing with your food and with the universe" (Bodio 1996, 230). The catching and the releasing are elemental structures of both our lives and the world we inhabit.[14] Both are "goddamned serious," as Thomas McGuane puts it (McGuane 1982, 232). One only has to see an angler, fishing a catch and release stream, unhook a fine fish as it flops on the rocks and then kick it back into the water, to know that some people have no sense of the serious and beautiful things in life, and no respect for life itself.[15] One can catch and release a beautiful fish with respect, and be on one's way.

REVERBERATIONS

> *Fishing for pleasure is quite ancient. Indeed, it can be traced back to the beginnings of orderly record keeping. Among the Chinese, the Greeks, the Romans, and later Europeans, fishing seems to have mattered and to have produced a certain giddiness, a theft of time in the sort of trickster's way we on this continent associate with the coyote, elsewhere the fox. Mimicking religion, angling sets out the steady movements of prayer meant to end in the vision, a captured fish. I wonder if the thing captured is the mercurial bond with nature, the need for which lies in greater or lesser measure in all of us* (McGuane 2000, 23).

In the summer of 1999 I took what outfitter Terry Search billed (accurately) as "the trip of a lifetime" into the remote Thorofare region of southeastern Yellowstone Park. It was late July, and although the main spawn of cutthroat trout up from Yellowstone Lake was over, Beaverdam Creek still had a lot of fish. I quickly realized that if I put my mind to it, I could probably catch and release two hundred fish in one day. It was midmorning when I deliberately slowed down. I watched the bald eagle watching me and the fish. I investigated the variety of small rocks and pebbles that had been washed out of the mountains to the east, including petrified wood from forests that grew millions of years ago. I caught and released some fish.[16] I watched fish rising to the sporadic hatches of gray drake mayflies. I spent more time watching forty cutthroats trying some late spawning in a long riffle, wishing them all success as the females dug redds and the males maneuvered for position. I had a wonderful day of varied experiences in one of the wildest parts of the lower forty-eight states. But the trout, those beautiful Yellowstone cutthroat trout (*Onchorhynchus clarki bouvieri*), now endangered by the lake trout some ignorant angler surreptitiously introduced into Yellowstone Lake about fifteen years ago, they were the constant *numenon*, the center around which the landscape came most intensely alive, and the rod in my hand was what most deeply connected me to that landscape, even when I was not using it.

There are many *numena* of the various landscapes we visit, live in, and love. One of Aldo Leopold's concerns in *A Sand County Almanac* was to sharpen our appreciation of them. This is the reason why Leopold the woodcock hunter wrote the beautiful little essay on the mating flights of woodcock in spring, "dancing in the sunset sky" (Leopold, 34). By the same token, there are many different practices in which these *numena* can make different landscapes come alive. Bird watching, encountering wolves in Yellowstone, hiking through grizzly country, gardening, the list can go on and on. Just seeing a tuft of grizzly hair caught in the bark of a tree beside the trail, or finding a pile of grizzly scat with a baby elk's undigested hoof surrounded by undigested elk hair, changed and charged my experience of Yellowstone, though I already knew, of course, that the bears were around. The more impact these activities have on the animals and systems involved, the more we have to think about why these practices are justified, especially in this world in which corners of wildness, to say nothing of so-called wilderness, are increasingly managed and tamed.

One crucial test is that of responsible stewardship. Do our practices, especially those which grant our lives genuine richness and put us in touch with the wild in ourselves in ways dominion can never achieve, enhance and protect the integrity of these *numena* and the biotic systems of which they are a part? At the end of several years of trying to think about these phenomena, I suggest that this is the question we have to answer. And as is so often the case, we find that Aldo Leopold has already given deep thought to the question. His famous sentences, "A thing is right when it tends to preserve the integrity, stability, and beauty of the biotic community. It is wrong when it tends otherwise" (Leopold 1949, 224–225) address one side of the question, the side that concerns the integrity and health of biotic communities, and our respect for them. The other side of the question concerns the individuals, human and nonhuman, who live in these communities. Here his evocation of the experience of the *numena* of the world plays the crucial role, for it is the cultivation of that experience that will guide our inquiry into the integrity of our practices that allow these *numena* to appear, in one way or another. Here individual paths will diverge. One person will be vegetarian; one will eat only the meat of wild animals he or she has killed; one will catch and eat fish; others will catch and release fish. I conclude that Albert Schweitzer was right both in his claim that there are compelling moral principles and in his claim that there is an irreducible subjectivity to our specific moral decisions when they are made in the framework of such principles (cf. Part II).

Approaching these problems from the point of view of animal rights or animal liberation oversimplifies the field, leaving both the animals and humans mere shadows of their true selves. Only such abstract thinking can lead one to conclude that it is our obligation to stay out of the world of animals, since such a conclusion does justice neither to the reality of human experience nor to the reality of animal life.

Our obligations, be it to other human beings or to nature, are always the correlates of our sense of self. But our sense of self is also a function of our experience in and of the world of which we are a part. The deeper our experience of that world is, the deeper our knowledge of ourselves has the potential to be, and the deeper our obligations become. What is purely and simply unacceptable is the claim that we have a general obligation not to experience, where true experience involves interaction and involvement, always necessarily against the background of appropriation. It is only right that the beautiful creatures of the world play deep and complicated roles in our lives. And it is only right that the world is the world in and from which we live—both on the level of organic existence and on the level of the meaning that permeates our lives. The question is how we can live with the fullest respect both for ourselves and for the world and all the creatures that find their home in it.

We live in a culture most of whose power centers—be they business or government (a distinction that is currently hard to draw)—have little respect for integrity and wholeness—be it the integrity and wholeness of self, the integrity and wholeness of family and community, the integrity and wholeness of a wild animal, or the integrity and wholeness of an ecosystem—and we need more than ever to cultivate experiences and practices that connect us with integrity and wholeness. Such connection is the essence of respect for nature. Hunting is, or can be, one such practice. In hunting one can experience the integrity and wholeness of both a system and an individual animal as part of the system, even as the hunter, as part of the system, attempts to appropriate, to kill and eat, that animal. The entire experience can be an encounter with, an expression of, and a contribution to the integrity and wholeness of the natural world.

The practice of catch and release fishing is most properly based on respect for the integrity of ecosystems and populations that are subjected to the pressures of human use and exploitation. Embedded in this practice is a specific respect for the individual fish one attempts to catch and then releases. This respect is embodied in the constraints the intent to release the fish puts on the methods and tackle used (bait such as worms tends to be swallowed more deeply than artificial flies, increasing the danger of serious harm to the fish), in the brief moments of pleasure we take in the beauty of the fish before we release them, and in the exquisitely gentle handling of the fish before it is released.[17] In such practices we gain a sense of the ways in which we can "fit well with" the land (Lopez 1987, 297) under changed historical circumstances, a sense of what the "new contracts" Lopez calls for might be. As such it is a discipline we impose on ourselves, not because it removes the taint of killing, but because it combines accepting responsibility for the consequences of our actions with contact with the *numena* that light up the world and our lives in it.

APPENDIX

FAITH, REASON, AND ANIMAL WELFARE

Periodically a book comes along that can remind its intended readers of who they are and aspire to be. Such books can challenge us unexpectedly, simply by appealing to our often unreflective religious and moral commitments. They can be noteworthy both for who the author is (a surprise) and for the way the author, in preaching to the converted, calls upon them to change their lives in ways that may surprise them. An argument that comes from a surprising corner—one that turns out to be our own!—may have the power to shake up our set ways of thinking, because we cannot write it off as being just what one would expect from, say, a conservative Republican.

Gordon K. Durnil's *The Making of a Conservative Environmentalist* (1995) is one such book. A conservative Republican who had been Indiana Republican State Chairman and member of the Republican National Committee in the 1980s, Durnil was rewarded by President George H. Bush with an appointment to the International Joint Commission, the U.S.-Canadian organization charged with maintaining the quality of the environment in the Great Lakes region. Given the studied lack of interest in the environment shown by the first Bush administration, it is hardly surprising that up to the time of his appointment in 1989 Durnil had little direct experience with or interest in environmental issues. But Gordon Durnil is one of those rare individuals who is capable of absorbing new information (in this case the reports by the staff of the Commission) and drawing practical conclusions that really are supported by that information along with his (in this case conservative) moral principles. What he discovered was that preservation, restoration, and, above all, prevention, are deeply conservative principles. While he remains committed to capitalism as an economic system and to a deep respect for individual liberty, he came to see that "the goal of environmental protection cannot be divorced from morality" (Durnil, x). Environmental protection cannot be left to the market for the simple reason that the market, as such, is no respecter of moral values. This is a deeply conservative principle, though few "conservatives" today give it more than lip service.

Durnil's book received some positive reviews and then, as far as I can see, it simply disappeared. There is little evidence that anyone in the administration of George W. Bush has read, much less digested, this expression of moral discovery and courage. The second Bush administration is, if anything, more committed to the identity of corporate and public interest than was the first. But the second Bush administration now has its own lonely voice crying in the wilderness—Matthew Scully.

Scully is a conservative Republican, and was a speechwriter for President George W. Bush from January 2001 to June 2002. He resigned this position just before publishing *Dominion: The Power of Man, the Suffering of Animals, and the Call to Mercy*, which challenges many attitudes concerning animal welfare that are widely shared by conservatives and Republicans, and by Americans of many political persuasions.

Targeting canned trophy hunts, commercial whaling, and industrial livestock production, Scully is polemical in the best sense of the word, appealing to what he assumes is a background shared by his reader, and then pushing it relentlessly. Most generally, this background is monotheistic religion, specifically Catholicism in Scully's case, but reaching out to embrace anyone rooted in the Torah of Judaism and/or the Old and New Testaments of Christianity.[1] "Judaism gave the world monotheism and a vision of the God who loves and cares for each person, and with that vision a view of the creatures as individual beings also known by Him, sharing with man not only in the earth's bounties but also—a still more intimate bond—in its punishments and suffering . . . The God of Israel delights in all that He has made. All creatures sing their Creator's praises, and are dear to Him for their own sakes" (Scully, 92). In *Genesis*, "the animals, too, were sent forth with a 'blessing' of their own" (*Ibid*, 11), and in *The Gospel According to Luke* Jesus is quoted as saying that though sparrows are sold in the market, "not one of them is forgotten before God" (*Luke* 12: 6–7; quoted in Scully, 95). Given these expressions of divine concern for animals, Scully asks how we can be thoughtless and cruel in our relations to them, treating animals as mere objects to be bought and sold to the highest bidder, and dealt with in any way that is economically efficient.

Most of the Judeo-Christian tradition has emphasized the rights that come with the dominion granted human beings in *Genesis.* Scully calls his readers back to their corresponding responsibilities, working to confront his reader with a very basic either/or:

> *If animals are just commodities, then we are just consumers, with no greater good than material pleasure and no higher law than appetite. And if there is a God and they are His creatures, not ours, then there is indeed a higher law regarding their care and we must answer to it—not just when it suits us, not just when we feel the spirit upon us, and not just when it's cost-efficient, but always* (*Ibid,* 308).

This either/or will surprise a utilitarian, of course, but at various points in the book Scully gives the utilitarianism of Peter Singer a pretty hard going-over.

Scully has three lines of argument against Singer: First, on a pragmatic level Scully argues that Singer's rejection of religion "surrenders far more ground than it gains, and from a strategic standpoint, at least here in America, it is worth noting that no moral cause ever got very far that could not speak to religious conviction, drawing on the deeper sensibilities that guide public opinion even in our more secular era" (*Ibid*, 13). Second, Scully argues that Singer's position on infanticide is properly based on his basic utilitarian principle and is itself immoral. It follows that Singer's basic principle has to be rejected. Given Scully's chosen audience, this is a pretty conclusive argument. Finally, Scully argues that the reason Singer's position produces such morally unacceptable results is that Singer has no basis on which he can put forth a moral principle in the first place.

> *Often his theories are not an appeal to morality or justice in any recognizable form, but a redefinition of justice. He requires of his readers, not that they apply common standards of morality, but that they accept an entirely new set of standards, and indeed a new and improved set of commandments produced* ex nihilo *from the mind of a modern intellectual (Ibid, 327).*

The problem is that for most readers there seems little reason to recognize these new standards as *moral* standards. Scully thus claims his traditionalism as a virtue. But *Dominion* is not written only for those who share Scully's monotheism of The Book. A second, subsidiary line of argument is also addressed to anyone who approves of our laws forbidding at least certain kinds of cruelty to animals.

> *Having granted some protections to some animals, we are constantly confronted with the logic of our own laws, troubled by perfectly rational connections between the random or 'wanton' acts of cruelty the law forbids and the systematic, institutional cruelties it still permits: If this animal is to be protected, why not that identical one, too? If it is cruelty to confine or mistreat a dog, a cat, or even a pet lamb or pig, why is it not cruelty to confine and mistreat millions of equally sensitive animals at Smithfield, IBP, ConAgra, and other such places? When we speak of the unavoidable severity of livestock production or laboratory experiments or trapping, and so on, just how rigorously are we defining 'unavoidable'?* (*Ibid*, 297).
>
> *We have, in short, the standard of internal consistency. We need not agree on everything to identify specific contradictions and falsehoods. You can operate by one or another of these basic assumptions [animals are commodities OR animals are deserving of moral consideration], but you cannot operate by a mix of them—here acting in blindly amoral fashion and there in the spirit of religious piety, here extending kindness and there spreading terror, here treating animals with respect and there treating them like refuse* (*Ibid*, 308–309).

In developing his argument, Scully tries to steer a course between the extremes of sentimentality[2] on the one hand and a "grim realism" that sees animals as mere objects to be disposed of as we see fit on the other. Both extremes exhibit a kind of myopia, in which the animals themselves disappear in our self-indulgent attitudes toward them.

> *As sentimentality towards animals can be overindulged, so, too, can grim realism, seeing only the things we want in animals and not the animals themselves. They do us a service if only by inspiring now and then a sense of wonder and humil-*

> *ity, for if not even a sparrow falls without His knowing then we are not too important to notice it ourselves* (*Ibid,* 2).

The appeal to "His knowing" is not an expendable metaphor here. Scully views animals as having a place and purpose in the world that is willed by God and, for this reason, something we must attend to. "Whatever measure of happiness their Creator intended for them, it is not something to be taken lightly by us, not to be withdrawn from them wantonly or capriciously" (*Ibid*). The fact that we live in God's creation means that we must attend to animals' "lives and place and purpose in the world" (*Ibid*).

We fail to live up to God's evident will when we treat animals as mere things to be dealt with in whatever way most benefits us. This does not mean that animals cannot legitimately be viewed in terms of what is beneficial and harmful to us, or that we cannot act in terms of that benefit and harm. "My point is that when you look at a rabbit and can see only a pest, or vermin, or a meal, or a commodity, or a laboratory subject, you aren't seeing the rabbit anymore. You are seeing only yourself and the schemes and appetites we bring to the world—seeing, come to think of it, like an animal instead of as a moral being with moral vision" (*Ibid*, 3). Scully is calling on us to see the rabbit itself even as it is a nuisance, since to see *only a pest* is to see *only yourself*. In both contexts, the key word is "only."

"Commodity" is the most important concept in Scully's list of a few of the ways in which we can fail to see the rabbit ("only a pest, or vermin, or a meal, or a commodity, or a laboratory subject"). As Scully's book unfolds it becomes clear that it is the basic category of his analysis, with all the others taking their place by falling under it. Just as sentimentality closes its eyes to the organic reality of animal lives, so the most institutionalized form of grim realism in our world is the relentless commodification that produces canned hunting and factory farms.

But to see something as a potential meal is not necessarily to see it as only a meal or as only a commodity. Scully, who is a vegetarian, agrees that Robert F. Kennedy, Jr.'s decision to eat only meat that comes from small farms that treat animals humanely is a "decent compromise" (*Ibid*, 316). To see the rabbit as a meal and to eat it is not what Scully objects to, at least not in this book.[3] The problem arises when we view the rabbit (or, more relevantly, the veal calf) *only* as a meal and thus *only* as a commodity, to be produced in as efficient a manner as possible.

As Albert Schweitzer recognized, the real source of much evil lies in thoughtlessness, and one of the most insidious aspects of commodification is the way it promotes thoughtlessness as it permeates everyday life, becoming invisible in myriad ways. Scully is acutely aware of this, and one of the threads that runs through *Dominion* is his criticism of the assumption that the outcome of individual decisions made in a free market framework ("consumer driven") is by definition "rational" and that price is a proper expression of value. Again and again he shows how easily the market becomes God or God the market.

A careful reading of Scully's chapters on trophy hunting and industrial farming lends support to the following generalizations. In a dynamic economic system, commodification at any level tends to produce greater or intensified commodification. As a corollary: Commodification is a certain kind of thoughtlessness, and unless it is carefully limited by morality and/or law, the process of commodification produces more and more thoughtlessness. Commodification tends to extinguish any form of thought or experience other than the technical thought of efficiency in production, selling, and buying. It is thus an institutionalized form of the "only." Any room for other forms of thought and value will tend to be squeezed out by the process of commodification itself. All noneconomic value tends to yield to that of efficiency, and does so in ways that tend to be invisible to us as producers and consumers. As Scully recognizes, when we treat animals as only commodities, God is dead in that part of our lives. "What the free market touches, it blesses. If the profit incentive points that way, why then that must be the right thing to do" (*Ibid*, 125).

If one accepts this critique of the commodity, it becomes clear that there are two revolutionary positions with regard to commodification, and a middle ground that is shared (and fought over) by liberals and conservatives alike. Whereas the Marxist position seeks the general overthrow of markets and commodities, the free market position affirms the process in which all public value is progressively commodified. The latter position cannot be called conservative in any legitimate sense of the word; it is rather a doctrine of permanent revolution on a pervasive level that Trotsky could not have even imagined. However, if one rejects viewing markets as either God or the devil, as either the key to heaven on earth or as the enslaver that must be completely annihilated by any means necessary, one can view them rather as one human institution, as one expression of human freedom among others. Then the question becomes *how* we are to live with commodification. This is where liberals and conservatives often disagree with one another to the point that their shared assumption, namely that there are crucial choices to be made here, becomes invisible. The ability of corporate interests to control the agenda of the supposedly conservative Republican Party is only one sign of this. But as Scully notes, there is a lot of ground for agreement between liberals and conservatives when it comes to the way animals are treated in our culture, since both liberals and conservatives agree that the question as to how we are to live with markets and commodities is a substantive and unavoidable issue.

Against the background of his critique of the commodity, Scully delights in taking on conservative pundits with whom he generally agrees, but who have a blind spot about the miserable reality of commercial trophy hunting and factory-style raising of domesticated animals. He relentlessly shows how conservative after conservative allows any sense of duty to or even a minimum of consideration for the animal members of God's creation to be subverted by the relentless drive for profits. "How low we can sink when, in the human mind, duty and profit seem suddenly and miraculously to speak as one" (*Ibid*, 271). His analysis of "The Prosperity Bible" should not be missed.

Scully's method is to confront his Judeo-Christian principles with the harsh reality of trophy hunting, commercial whaling, and industrial farming. In the chapter on trophy hunting Scully recounts attending a convention of Safari Club International. By simply describing the marketing of canned hunts, the rampant egotism of the Club's complex system of awards for killing ever more species of trophy animals, and the appalling crassness of the "hunting" videos that are ever-present, Scully simply lets the intrepid hunters, including retired General Normal Schwarzkopf as keynote speaker, hang themselves out to dry. In the chapter on industrial farming Scully recounts his visits to pig-raising facilities run by Smithfield Foods, Inc. He simply describes the conditions he finds, and allows representatives of the corporation to speak. Once again, they hang themselves to any reader who has not decided in advance that pigs are just resources to be used in the most efficient ("consumer driven") way possible, and that they are either not sentient to begin with or their suffering simply does not matter ("mere pain"). In both cases, the simple incapacity and unwillingness to think about what they are doing is appalling. (I personally find the chapter on commercial whaling less compelling, though I am sympathetic. Here he goes well beyond preaching to the converted, and for this reason his polemic becomes shrill.)

In his analysis of the consequences of progressive commodification, Scully has a sharp analytical tool, but he sometimes draws conclusions that go beyond what his analysis has shown. The chapter on trophy hunting is an example of this. Scully does a wonderful job of allowing the spokespersons of the trophy animal industry (from both the supply side and the demand side) to reveal the despicable reality that is the result of their relentless objectification and commodification of wild and "wild" animals. As he condemns the commodification of wild animals as trophies offered for a price, he assumes that this extends to all contemporary hunting, but this is an overgeneralization. One can begin to see this by comparing Scully's fine treatment of Safari Club International with Ted Kerasote's equally fine treatment of international trophy hunters in his *Bloodties: Nature, Culture, and the Hunt*. Kerasote's strategy is very similar to Scully's: he accompanies wealthy trophy hunters on a trip to Siberia, and allows their words and actions to speak for themselves. But unlike Scully, Kerasote does not assume that the moral bankruptcy of commercialized trophy hunting demonstrates the moral bankruptcy of all contemporary hunting. Instead, Kerasote also travels with indigenous hunters in Greenland, letting the subsistence hunters speak for themselves, and he goes on to speak for himself as someone who hunts the elk of the Grand Tetons, and does so as a part of the ecosystem he and the elk share.

Scully never mentions authors such as Kerasote, and is instead satisfied with quoting the shallow spirituality of James A. Swan's *In Defense of Hunting*. If he had tested his general opposition to hunting against Kerasote's account, I suspect that he would have to concede that Kerasote's way of living and hunting is at least as much of a "decent compromise" (*Ibid*, 316) as eating only meat from small and humane farms. (But the fact remains that for Scully it is a compromise, however decent, suggesting that it is the vegetarian who occupies the moral high ground, whose moral aspirations [cf. *Ibid*, 311] are higher. Kerasote effectively challenges this.)

Given Scully's very effective strategy of appealing to a shared faith in God and a shared tradition concerning God's relationship to the world, it is understandable that he decides to "leave the theories to the theorists" (*Ibid*, 299). Yet he also sees the need at least to sketch what he calls the "rational grounds" that underlie all moral claims, beyond all sacred texts and all philosophical theorizing. Here he appeals to the tradition of natural law.

> *This key insight is that all moral truth arises from the nature of things, true in themselves and in crucial respects accessible to reason. Every being has a nature, and that nature defines the ends and ultimate good for which it exists. In discerning these purposes we perceive what that being is, what it can do, what it must do to find its completion and fulfillment, and therefore what its moral interests are and how they may be advanced or hindered. Suddenly all is not arbitrary and we have a fixed point of reference, an intelligent basis for calling one thing good and another bad. That which advances a being onward toward its natural fulfillment is good. That which frustrates or perverts its natural development is bad* (*Ibid*, 299–300).

Given the initial emphasis on "the nature of things," one could say that "natural law" is the law of respect: the injunction of natural law is to *respect the natures of things*. How is this to be understood? As it is stated here, natural law has to do only with the natures of things, not with the transcendent ground of their existence and its meaning. The good and the purposes for which a being exists are *its* good and *its* purposes. God may have some end in view as the ground of all existence, but that is not the point here. The appeal to reason is in this sense philosophically agnostic. "It [natural law] provides the only rational grounds I know of for claiming any one thing better than another, without reliance on religious belief or intuition or the constructs of theory" (*Ibid*, 301).

As a first approximation, Scully's position seems to be the following:

1. Every being has a good of its own, rooted in its nature.
2. A being's good is morally relevant.
3. It is (morally) good to advance a being's interests, (morally) bad to frustrate or pervert a being's interests.

Scully is casting his net very widely here, perhaps too widely. If *every* being has a nature and thus a good, this includes not just animals, but also plants, and indeed *every*thing. This is not nonsensical, and not without precedent: Albert Schweitzer's ethics of reverence for life led the great humanist to view everything in our world as an expression of an infinite will to live (cf. Part II). The Spanish philosopher José Ortega y Gasset argued that we have obligations to rocks such as the "venerable rocks of the mountain ranges" (Ortega y Gassett, 89). More recently the philosopher Paul Taylor has argued that plants have a good of their own and are thus deserving of moral consideration (cf. Part III). Given the focus of Scully's book, it makes sense that he does not investigate this further. But there are questions and problems here. The problems come in the moves from 1. to 2. and from 2. to 3.

From Scully's perspective, the problem in the move from 1. to 2. is, I think, the less pressing of the two. A series of philosophers over the past thirty years or so—including Tom Regan (cf. Part I), Kenneth Goodpaster, Paul Taylor (cf. Part III), and others—have argued that if a being has a good, such that it makes sense to speak of respecting that good, then we have an obligation to respect that good, that is, to take it into account in moral deliberation. The mere fact that there is a good to respect is all that is needed to establish what is often called "moral considerability." But there is something less than universal agreement here, to say the least, so it is hard to avoid speaking of something like a moral intuition at work. When natural law is reembedded in Scully's Judeo-Christian monotheistic tradition, of course, the connection receives religious sanction. It would be possible simply to fall back to the weaker claim that having a good makes a being something that *can* be given moral consideration, the imperative for actual consideration coming from other, perhaps religious, sources. Again, given the focus of Scully's book, it makes sense that he does not investigate this further.

The move from 2. to 3. is more difficult. There are two layers of meaning in 3—First, the position is that since whatever advances the natural fulfillment of a being is good in the sense of being good *for* it, a moral agent's action that advances the natural fulfillment of another being is good in a higher sense—namely *morally* good. But it is one thing to argue that if a being has a good, then that good is relevant to my moral deliberations, and something quite different to say that my action is morally good or morally bad when it advances or frustrates the natural development of that being. But Scully also seems to be saying something that goes beyond this formulation. In writing, "that which advances a being onward toward its natural fulfillment is good," Scully seems to be saying that whatever advances the natural fulfillment of a being is *morally* good, and whatever frustrates that natural fulfillment is *morally* bad, is *evil.* Not just human actions, any event can be found to be morally good or morally bad.

What reason does Scully have for taking these positions? Here we need to take a closer look at a phrase that is easy to pass over without paying proper attention. What does it mean when Scully writes, "Every being has a nature, and that nature defines the ends and ultimate good *for which it exists*" (*Ibid*, 300; my emphasis)? The phrase "for which it exists" is ambiguous.

On the one hand, it can mean that this good is *its* good, a matter of what is good *for it.* When one speaks of "the ends and ultimate good for which it exists" in this sense, one is saying that this being has a good of its own, and that we can evaluate events in the world as being good, bad, or indifferent with respect to the well-being of that organism. But to say that an event is bad for, say, an elk (the elk is killed by wolves) is not to say that anything that is bad without qualification (as opposed to being bad *for the elk*) has occurred, much less something immoral or evil. Indeed, the same event is good . . . for the wolves. There is nothing about the natures of the elk, considered in isolation, and of the wolves, considered in isolation, that will allow us to say that this event is itself either good or bad in an unqualified sense.

But there is a second sense in which we can speak of a good for which an animal exists. Here we speak of the ground of that being's existence, the very meaning of its existence. For example, we could follow the Western religious and metaphysical tradition of unbridled anthropocentrism in saying that animals exist for the use of human beings. In this case the animals' instrumental value to humans is the very meaning of the animals' existence, the ultimate answer to the question as to why these animals exist at all. Their goodness is that they are good for . . . us. Scully rejects this tradition, so when he speaks of the good for which an animal exists, it is the intrinsic goodness of the animal, as it were the intrinsic goodness of the animal's good, which is at issue. (From the religious perspective, the expression of this intrinsic good is that God blesses the animals and cares about them.) One might hold that if animals are intrinsically good, then they are not mere instruments to be used in any way we see fit. Our use has to be compatible with recognition of and respect for their intrinsic goodness. But Scully seems to go farther and say that, strictly speaking, respect for an animal's intrinsic goodness is incompatible with *any* use that "frustrates or perverts its natural development." All such use is intrinsically bad, evil. This would permit many, but not all, pets, but would condemn raising or killing animals, any animals, for food.[4]

How can we determine which way to go here? It might seem that there is really only one alternative here: if "that which frustrates or perverts [an animal's] natural development is bad" (*Ibid*, 300), then such action is immoral. If this is the case, then Scully is simply recognizing the nature of things when he identifies "the ends and ultimate good for which it exists" with an animal's "moral interests." But there is an alternative here. If we look at, say, elk from the perspective of evolutionary ecology, we can see that elk co-evolved with wolves over many thousands of years. Many of the elk's characteristics developed as they did because elk evolved as prey animals for wolves. The same thing can be said about wolves as predators. Something similar can be said about beaver, which tend to disappear in some habitats when wolves are removed, because elk then over-browse the willow saplings that are crucial to the beaver. The ends and ultimate good of the elk, both individually and as a population, are intricately intertwined with the good of wolves and beaver, as well as the good of coyotes, pronghorn, rodents, raptors, and even willows, if the good of trees is taken into consideration. In speaking about the "ends and ultimate good for which elk exist" in an intact ecosystem, we cannot divorce this good from the ends and ultimate good for which wolves and beaver exist. If these creatures are good, so are their interrelationships, without which the creatures become mere abstractions. From the religious perspective, if all creatures are dear to God for their own sakes (cf. *Ibid*, 92), then it must be the case that these interrelationships also receive God's blessing.

This perspective is helpful in getting the kind of purchase in the real world that natural law requires if it is to give real moral guidance. Prior to all moral theorizing,

> *[Natural law] serves as a kind of anti-theory, an understanding of natural proprieties, an acceptance of things as they are prior to man's attempts to intellectualize them and make them his own. It compels us by reason to perceive purposes and goods beyond our own desires and decrees, to heed natural boundaries, to respect and live within an order of which*

> *we are a part but not the center. It asserts what the philosophers call a teleological view of a moral universe with a detectable structure, direction, and broad design beyond our power to alter or escape.* (*Ibid,* 301)

The question at this point is just how we are to understand the "natural proprieties" and the "moral boundaries" made visible by natural law. Just what is the *telos* of this moral universe? In considering the natures of things with an eye to discovering "natural proprieties" and "natural boundaries," one might consider the long legs, huge feet, powerful jaws and teeth, and so on of the wolf and recognize that the good of the wolf involves hunting animals such as elk as food. Wolves find their natural fulfillment in living in packs, hunting elk, reproducing, nurturing pups, and so on. When they are successful, this is good: not just good for the wolves, but good. From a religious perspective, in watching wild wolves, one is viewing the glory of God's creation. By the same token, we might consider the strength and quickness, the alertness, the shape of the teeth, the rumen, and so on, of the elk, and recognize that the good of the elk involves browsing, being alert for predators, caring for their young, and so on. When they are successful, this is good: not just good for the elk, but good. From a religious perspective, in watching wild elk, one is viewing the glory of God's creation.

What, then, can we say about events that advance or frustrate the natural development of elk and wolf? That which advances the elk toward their natural fulfillment, for example, a mild winter, is good, in the first instance, *for them*; that which frustrates the elk, such as a hard winter, is bad . . . for them. But the same thing holds for the wolf, with the difference that a hard winter is good for wolves while in a mild winter the wolves have a hard time. One might recognize in these facts something of the nature of "an order of which we[5] are a part but not the center." When we view this universe teleologically, namely as an order of existence that is intrinsically good (a "moral universe"), in which these individuals have their place but in which they are not the center, we would recognize that an ecosystem in which both elk and wolves (and much more) can thrive is a healthy, good ecosystem. The good of the elk is bound up with the good of the wolf in such a way that what is bad for an elk can be good for a wolf, both of these embedded in the goodness of a world in which elk and wolves are possible.

Is a hard winter, beyond being good for wolves and bad for elk, itself good or bad? The answer would seem to be "good," since the climate, with all its variability, is part of the order that makes both elk and wolves possible. (The alternative answer would be "yes.") When Scully writes, "we can discern that they have each a place to fill and a purpose to live out as well" (*Ibid*, 302), that place is bound up with their niche in a larger whole. To ignore it is to reduce the animals to abstractions. From a religious perspective, in watching elk and wolf face the challenges posed by both harsh and mild winters, one is viewing the glory of God's creation. Finally, if one has the opportunity to watch a pack of wolves chasing elk, one is viewing the glory of God's creation, regardless of whether the wolves are successful or the elk escape. In watching a kill, one is viewing the glory of God's creation; in watching a successful escape, one is viewing the glory of God's creation.

All of this seems to fit perfectly with Scully's integration of natural law and his monotheistic religious orientation.

> *Add a little old-time religion and natural law has the further advantage of being an affirmation of life as good and purposeful, instead of treating life, in the way of so many contemporary philosophers, as some gnawing problem to be solved or escaped. It sees every life, ours and the lives around us, even in trial and sorrow as the gifts they are—no creature being slighted in being what it is, all exactly as they are meant to be.* (*Ibid*, 304)

And yet, in the final analysis, Scully does not accept this entire line of reasoning. His "teleological view of a moral universe with a detectable structure, direction, and broad design beyond our power to alter or escape" (*Ibid*, 301) does not include predation. The reason for this does not lie in natural law as he presents it, but rather in his own religious appropriation of natural law. It turns out that the way animals are is not in fact "exactly as they are meant to be. "Predation itself, the intrinsic evil in nature's design of creatures devouring and absorbing one another to survive, is among the hardest of all things to fathom. One falls back in the end on the idea that it was not God's design at all, that there lies a hope and expectation beyond creation's 'groan of travail,' as we are promised, not only mankind but all creatures delivered from our 'bondage to decay' (*Ibid*, 318).

From this perspective, predation is not God's will, but a consequence of the Fall (*Ibid*, 44), the result of human sinfulness.[6] This is not just religion's appropriation of natural law as Scully has presented it, it is rather religion's rebellion against natural law. From the religious point of view as Scully understands it, things are not "all exactly as they are meant to be" (*Ibid*, 304), since in central ways the world is "not God's design at all" (*Ibid*, 318). This has little to do with recognition of the animals' natures. So either it was a mistake to affirm the value and goodness of the intertwining of the lives of elk and wolves in the first place, or Scully's religious perspective simply rejects anything in natural law it finds incompatible.

Why does Scully arrive at this state of affairs, the religious point of view from which most of his book is written standing in contradiction to the natural law he puts forward as its rational core? The key to what is going on is found in a curious change in Scully's attitude toward science. For most of the book, Scully appeals to science for the knowledge of the world required for a proper moral orientation in the world. (This does not mean, of course, that science can discover what a proper moral orientation is.) "The scientific proposition 'animals suffer' is either true or false. It is not an opinion, but one way or the other an objective reality" (*Ibid*, 193). When some scientists and philosophers hold either that animals do not suffer or that their pain is so different from human pain as to be insignificant ("mere pain"), Scully has done enough homework to argue that this is bad science—science that has lost its scientific objectivity—and bad philosophy. And when he explains the fact that the most distinguished advocates of natural law, Aristotle and Aquinas, both held that animals are properly used as tools for human benefit, he writes that they "just didn't know as much about animals as we do" (*Ibid*, 302). Progress in scientific knowledge leads to progress in our understanding of natural law.

And yet, in the course of his religious appropriation of natural law Scully simply rejects an evolutionary perspective. Rather than limiting himself to arguing (as he does) against the modernized social Darwinist conclusions of writers such as Stephen Budiansky and rejecting them as bad science and bad philosophy of science, he simply places "the fundamentalist" (who argues for human dominion without any corresponding obligation) and "the evolutionist" (who sees human beings "as the product of blindly amoral evolution") as extremes that "speak a common language of power, appropriation, and consumption" (*Ibid*, 306). So rather than viewing the wolf's long legs from both the perspective of evolution (co-evolved along with elk in the course of becoming a predator that chases its prey) and the perspective of divine creation (seeing the goodness and beauty of God's creation in the long legs, in the graceful and efficient stride, *and* in the ability to be an efficient predator), Scully thinks that it is an either/or. It is hard to avoid the conclusion that at this point Scully falls prey to the sentimentalism he wanted to avoid.

Dominion is a book about the commodification of animals, not about our responsibilities to and for the broader natural environment. But beyond this necessary limitation in scope, Scully distances himself from both certain forms of environmentalism (cf. *Ibid*, 102) and, it would seem, "modern environmentalism" itself (*Ibid*, 130). (This may explain how Scully, who returned to writing speeches for President George W. Bush after publicizing his book, can be part of an administration that has relentlessly supported the commodification of the natural world in its most extreme, corporate, form.) But if my discussion of Scully's position on natural law has any validity, moral issues of animal welfare are part of the broader issue of the human relationship to the natural world, environmental ethics. The evils that flow from the progressive commodification of animals are of a piece with the evils that flow from the progressive commodification of the natural world. In both cases, morality cannot orient itself toward the market and the interests of corporations. "Free market" or "consumer driven" morality is no morality at all.

Dominion is a very important book. It should be read and taken to heart not only by the Jews and Christians to whom its arguments are primarily directed, but by everyone. I have taken issue with some of the philosophical underpinnings Scully offers, but this does not undermine the central argument of his book, which is that we as a culture tolerate the hideous and institutionalized abuse of animals, and that this immorality is hidden by a thoughtlessness that has its roots in the process of commodification. At the very center of our corporatist moral, economic, and political culture lies the heart of darkness. Matthew Scully calls us, all of us, back from our worship of the god of economic efficiency and economic rationality, the god of profits, and forward to becoming the selves we are committed to being.

NOTES

PREFACE

1. This meaning of the term is of fairly recent origin. In the *Historisches Wörterbuch der Philosophie*, Volume I, 1971, the term "*anthropozentrisch*," is dated to the second half of the nineteenth century. The context at that time was theological, not environmental or ethical, and the term's meaning varies according to author and context. Thus, one theologian contrasted the "theocentric world view of Christianity" with the "anthropocentric" world view of Rousseau, while the famous historian of philosophy, Wilhelm Windelband, wrote that the Christian view of the world has an "anthropocentric character," since Christianity puts humanity at the center of the (created) universe (Birkner, 379). The specifically environmental significance of the concept is not even mentioned. Today it would be unavoidable.

There have, of course, been counter-forces of various kinds. Pantheism is the most obvious. St. Francis of Assisi is often cited. Goethe wrote that the idea that human beings are the "final purpose" of nature is an expression of human vanity (Goethe, 82). Jeremy Bentham and Albert Schweitzer also denied that humans have superior moral status (cf. Parts I and II).

2. Lynne White, Jr. finds this text crucial for supporting his claim that "Christianity bears a huge burden of guilt" for the destruction of nature (White, 1206). But *Genesis* also has a second account of creation, in which God tells man "to cultivate and take care of [the garden of Eden]" (*Genesis* 2:15), which can be seen as the source of a duty to be the steward of creation, as opposed to a destructive and self-centered dominator (cf. Callicott 1981, 136–139; Dobel).

3. Aristotle presents a nontheologically-based argument to reach similar results: there is a natural hierarchy of higher and lower between soul and body, humans and animals, animals and plants, male and female, higher men and lower men (who are "by nature slaves") (Aristotle, 1254b5–20). Concerning animals, the principle is that "if nature makes nothing incomplete and nothing in vain, the inference must be that she has made all animals for the sake of man" (*Ibid*, 1256b20–22).

The Epicureans argued against eating meat, not out of respect for animals, but for reasons of personal purity and health (reported by Porphyry in *On Abstinence*, I.51.6–52.1; in Long and Sedley, 116).

The Stoics accepted Xenophon's theological view (cf. Cicero, *On the Nature of the Gods* 2.133; in Long and Sedley, 328). But Porphyry quotes Carneades' telling critique: "Now let anyone who finds this at all persuasive, and fitting for god, consider how he is going to reply to the following argument of Carneades. 'Every product of nature, when it achieves the natural end for which it was born, is benefited . . . But the pig has been born for the natural end of being slaughtered and eaten. When this happens to it, it achieves its natural end, and is benefited'" (Porphyry, *op. cit.*, 3.20.I, 3; in Long and Sedley, 329). Porphyry clearly considers this to be a *reductio ad absurdum*. (Thanks to my colleague Eric Brown for these references.)

The Victorian poet James Henry makes fun of anthropocentric vanity in his long poem "Man's Universal Hymn." One verse reads:

> *I love the Lord my God, for he*
> *Loves all his creatures tenderly,*
> *But more than all his creatures, me.*
> *He bids me from the dam's side tear*
> *The tender lambkin and not spare:-*
> *'Piteous though bleat the orphan'd dam,*
> *Túrn a deaf ear and dine on lamb'* (Henry, 36).

(Thanks to Guy Petzall for this reference.)

4. It is striking that Thoreau's attraction to vegetarianism is tied to an opposition between the soul and the body, and that in his writing vegetarianism is closely linked with the ideal of chastity. He begins the chapter of *Walden* entitled "Higher Laws" by affirming his "reverence" for both the "spiritual" and the "rank and savage" life, and he claims to love "the wild not less than the good" (Thoreau, 384). Thoreau approaches an attitude of biocentric egalitarianism when he writes, "No humane being, past the thoughtless age of boyhood, will wantonly murder any creature which holds its life by the same tenure that he does . . . I warn you, mothers, that my sympathies do not always make the usual phil*anthropic* distinctions" (*Ibid*, 386). Here there is a recognition of the inherent value of the hare, and Thoreau's rejection of the "phil*anthropic*" surely moves in the direction of the biocentric outlook, or of what A. O. Wilson calls "biophilia." But this recognition of inherent value is quickly tied to his distinction between the lower and the higher: not only is there "something essentially unclean about this diet [of meat] and all flesh . . ." (*Ibid*, 387), such fare is appetizing to "an animal to us, which awakens in proportion as our higher nature slumbers" (*Ibid*, 390). Sensuality, and "all sensuality is one," is in opposition to purity, and "all purity is one" (*Ibid*, 391). Vegetarianism, like chastity, is a matter more of self-purity than of respect for that which is other. This is a theme that will reappear in Parts II and III.

5. William Cronon has noted this counterpart relation in a different context. Cf. Cronon, 484.

6. I generally use the term "inherent worth" because it is the word Paul Taylor uses. As Taylor notes, his "inherent worth" is "essentially identical" to Tom Regan's "inherent value."

7. Personal communication concerning an earlier draft of this "Preface." The last two paragraphs were stimulated by Rolston's comments.

CHAPTER 1

1. An earlier version of this chapter was presented at St. Louis University and at the conference on "The History, Ethics and Philosophy of the Hunt," sponsored by Orion: The Hunter's Institute. Feedback from both audiences was very helpful.

2. Indeed, after reading this part of the book, Tom Regan suggested that Singer simply isn't interested in the idea of "respect" examined here, and that it might be best to drop Singer from the discussion entirely. Regan is right that Singer uses the "not as means to our ends" expression rhetorically for his own utilitarian purposes. From the perspective of Regan's animal rights theory, and indeed from my own perspective as it begins to unfold in this section, Singer's use of the phrase, like his use of the word "rights" in some early essays primarily for rhetorical effect, do not, strictly speaking, fit his theory. But I have chosen to include Singer here because he does make such rhetorical appeals.

3. This formula also fits the biocentric positions discussed in Parts II and III. Animal rights and animal liberation thought is distinguished from biocentric thought by the fact that the latter extends the recognition of inherent value (or moral considerability) to nonanimal organisms. Some forms of biocentric though also extend such recognition to species, communities, and ecosystems as well (cf. Part III).

4. The immediate source might be Paul Taylor's concept of "respect for nature."

5. In his comments to this section Tom Regan notes that there are limits to the "keep your distance" policy in that "we ought to defend animals in the face of human abuse, cruelty, etc."

6. Kant is clearly talking about adult human beings in the first instance, and would say that it is the nature of infants to develop into actually rational beings. He also qualifies inclusion of women.

7. Kant, of course, approaches the matter in terms of duty. But since the concept of "duty" or "obligation" can be defined in terms of that of "permissibility," this is simply a matter of two ways of saying the same thing.

8. Brian Luke recognizes this kind of "flaw" in the arguments of Singer and Regan (Luke 1996, 78–79), though he then goes in a very different direction than the one investigated here.

9. In his comments Tom Regan notes that this is not his view, and he is clearly correct. I am following out the implications of a Kantian position that Regan does not adopt. My goal here is not to refute Regan, but to develop an alternative.

10. It goes without saying that neither Singer, who is after all a utilitarian, nor Regan, who does not accept this version of Kantian ethics, would find this convincing. But that is not the point. What I am interested in is less a refutation of ethics by extension (if such is possible in this field) than in opening up a different line of thinking.

11. Claims that traditional Native-American, or more generally hunter-gatherer cultures, are more ecologically enlightened than modern industrial cultures are both common and controversial. At their worst, they are a new version of the image of the "noble savage."

There have been a series of attempts to look at Native-American attitudes toward nature while avoiding romanticism (cf. Booth and Jacobs; Buege; Callicott 1989a, 1989b; Callicott and Nelson; Jacobs; Nelson 1983, 1994; Vecsey). Perhaps the best study of the "myth that non-industrial cultures are ecologically benign" is Kay Milton's *Environmentalism and Cultural Theory*, chapter 4. More recently, Shepard Krech III, in his book *The Ecological Indian: Myth and History*, has investigated the recorded history of Native-American practices and compared them to modern conservation. His conclusion is that "the ecological Indian" is a myth. His analysis is important, and a good antidote to tendencies to romanticize. However, as he notes, the records he analyzes are those of post-contact Native American life, when trading with the Europeans had already begun to change the older traditions.

Some would simply reject Barry Lopez's claims as romantic nonsense, but his existential reflections have a firm anthropological foundation. Modern conservation is not the only point of view from which to investigate Native-American attitudes and practices. Richard Nelson, for example, examines Koyukon "ecological patterns and conservation practices" in their own terms (cf. Nelson 1983, 200–224), and not those of contemporary environmentalism. (Kay Milton has emphasized the importance of this distinction.) Lopez could easily be read as a more poetic gloss on Nelson's findings. One need not generalize in order to find the practices Nelson describes to be of interest precisely as human possibilities.

Finally, while I will make use of some ideas developed by Paul Shepard, it should be recognized that many of his ideas are more romantic than science-based, and some of the science he appeals to is now clearly dated and, at best, partial.

12. This quote is not taken out of its immediate context, but Thomas's book as a whole moves from this sense of hunting as "fun" to a much deeper sense of the meaning of hunting. I return to this in Part IV.

13. One could go farther and argue that once we have removed the predators of, for example, elk from an ecosystem, we have an *obligation* either to reintroduce the predators (as in Yellowstone) and/or to hunt the elk. Otherwise our very solicitude for the animals' welfare (as defined by animal rights/liberationist thought) will lead to their decline. This suggestion goes

beyond the notion of "obligatory management species" as developed by Gary Varner (cf. Varner, 101f.).

14. The selections from McGuane could be supplemented from reflective hunting literature almost indefinitely. The issue of the hunter's felt "remorse" and "guilt" is a topic deserving extended discussion (cf. Part IV).

CHAPTER 2

1. *The Philosophy of Civilization*, Schweitzer's main systematic work on the problem of ethics in the twentieth century, has not been served well by translators. The original translation of *Kultur und Ethik*, which is Part 2 of Schweitzer's unfinished *Kulturphilosophie*, by John Naish appeared in 1923. It was supplanted by a translation of both published parts under the title *The Philosophy of Civilization*, by C. T. Campion in 1929. It reappeared in 1945, this time with revisions of Part 2 by "L.M.R." (cf. "Reviser's Note," 67–69). But the translation remains somewhat stilted, at some places misleading, and at others simply confusing. While I have used this translation as my point of departure in quotations, each quote has been checked with the original, and when necessary the translation has been changed without special note.

The most glaring problem in Campion's translation, other than individual infelicities, is the translation of the German words "*Hingebung*" and "*Hingabe*". "*Hingebung*" is translated as "self-devotion," "altruism," "self-dedication," and "devotion," (cf. pages 296, 299, 305). The basic meaning is "devotion," and the translation "self-devotion" is very confusing and misleading in several contexts. But even this is an improvement over the Naish translation "self-sacrifice."

2. The next chapter will investigate Paul Taylor's principle of respect for nature as an adaptation of Schweitzer's principle of reverence to more contemporary environmental attitudes and concerns.

3. When Schweitzer is read by environmental philosophers, it is often only selected excerpts, with the result that the selected passages remain incomprehensible. An example is Tom Regan's short discussion of the shortcomings of Schweitzer's principle of reverence for life in *The Case for Animal Rights* (cf. Regan 1983, 241–242). When Schweitzer writes that the truly ethical person, the person who exhibits true reverence for life, "shatters no ice crystal that sparkles in the sun," Regan objects that "There is not clear sense in which ice crystals are 'alive' or exhibit 'will to live'" (Regan 1983, 242). (This phrase does not appear in the original German edition of 1923, nor in the revised Campion translation of 1949. It does appear in the Naish translation of 1923. Regan refers to the original Campion translation of 1929, but the passage he cites is identical with the Naish translation.) The problem with this is that Schweitzer has a carefully worked-out sense in which *everything* is an expression of "the mysterious will-to-live that is in all things" (Schweitzer 1923, 309). One may

reject his mysticism, but it should be noted that this is Schweitzer's position and that he offers reasons for it.

4. Much of *The Philosophy of Civilization* was written in the evening after his day's work in the hospital in Africa. In his little book on Schweitzer, the philosopher Oskar Kraus notes that the manuscript bears the notation "in great weariness" at numerous points (Kraus, 44). Schweitzer worked on the promised Part 3 of *The Philosophy of Civilization*, which was to be titled "The World-view of Reverence for Life," until 1945, at which point he abandoned it. A sketch of its contents, and of the difficulties Schweitzer encountered as he worked on it, can be found in Claus Günzler, *Albert Schweitzer: Einführung in sein Denken*, Part V.

5. I point out other parallels between Schweitzer and Kierkegaard below. It is striking that Schweitzer never mentions Kierkegaard.

6. "Life-philosophy" [*Lebensphilosophie*] should be carefully distinguished from "philosophy of life" (*Philosophie des Lebens*].

7. Schweitzer's onetime teacher Georg Simmel belongs to this group.

8. The German word "*Geist*" has a broader meaning than the English words "spirit" and "spiritual." It encompasses not only the religious sphere but also culture (which Hegel calls "objective spirit") as a sphere of shared values. Thus, Schweitzer's "higher spirituality" involves development of a higher culture, but such a culture will also necessarily be spiritual.

9. Nietzsche's infamous "*Übermensch*," which I generally prefer to translate as "the beyond-human."

10. This critique of Nietzsche echoes Søren Kierkegaard's critique of the aesthetic approach to living in *Either/Or*. In essence, Schweitzer accuses Nietzsche of making life meaningless under the guise of raising life to its highest power. Kierkegaard's word for this is "despair." For Schweitzer's more general statement of this "Kierkegaardian" position, cf. Schweitzer 1923, 278–281.

11. Schweitzer's interpretation of Nietzsche is, I think, very weak. I return to this below.

12. An ecologically-oriented version of this idea is developed by Arne Naess in his "ecosophy T" as his personal ecological wisdom underlying "deep ecology" (cf. Naess, 1988).

13. Schweitzer writes here of returning to "*einem elementaren Philosophieren.*" The word "*elementar*" appears often, and Campion tends to translates it, as he does this passage, as "elemental." I have avoided this, because "elementary" seems closer to what Schweitzer has in mind in his usage, which has none of the (very Nietzschean) overtones the word has in the work of Ernst Jünger, which was beginning to take shape as Schweitzer wrote his *Kulturphilosophie*. In Schweitzer, "elementary" always means "existential."

14. When he begins to develop his own approach to ethics in chapter 23 of Part II of *The Philosophy of Civilization*, Schweitzer often makes use of what might be called the "universal 'I.'" While he does use the plural "we" at times, the "I," which is not restricted in meaning to "I, Albert Schweitzer," is crucial to his elaboration of an experience and reflection that

can only be each individual's experience of and response to that individual's own will to live. In presenting and interpreting Schweitzer's philosophy, I (as writer) will repeatedly make use of this universal "I." It is hoped that attention to context will make it clear when the "I" is universal, and when it refers to the writer as writer.

15. It is easy to hear the echo of the opening paragraph of the "Eulogy On Abraham" in Kierkegaard's *Fear and Trembling*.

16. Kierkegaard's esthete presents the esthetic life as being rooted in immediacy, as the attempt to choose immediacy and live it, while Schweitzer presents his version of the esthetic life here as a *result* of pessimism, that is, as being already a form of pessimism rather than inevitably producing it. But the difference may be more apparent than real. In the text on *Don Giovanni* ("The Immediate Erotic States OR The Musical-Erotic"), Kierkegaard's esthete writes that the immediate upsurge of desire and satisfaction that is Giovanni is "born in anxiety," that "Don Giovanni himself is this anxiety, but this anxiety is his energy" (Kierkegaard 1843a, Vol. I, 129). The sensuous upsurge of life is not as immediate at it initially seems.

17. Schweitzer's text in this section mimics the repetitious and aimless to-and-fro of such a life caught between its natural will to live and the failure of the world to confirm this affirmation of life.

18. It is also very close to Nietzsche's will to power, although the differences are crucial.

19. Aristotle notes in the *Nichomachean Ethics* that one should not demand more rigor and precision than the subject matter allows. The unstated corollary is that one should demand every bit of rigor that the subject matter does allow.

20. It is interesting that in his translation, Campion uses "pity" in the context of Schopenhauer's ethics (Schweitzer 1923, 239), and "compassion" when Schweitzer is developing his own thinking, although the German has "*Mitleid*" in both places. There is a certain justice to this, because Schopenhauer's ethics is based on a negation of life while Schweitzer's is an affirmation of life. Schopenhauer's "pity" may really be an essentially different passion than Schweitzer's "compassion." I return to this below.

21. Schweitzer could recognize this to be the limited truth of Nietzsche's thought of the will to power.

22. It is tempting to read this "*eins werden*" (literally, "becoming one," which along with "union [*Einswerden*]" is Campion's translation) as something much more radical than the "fellowship" and "solidarity" he speaks of a few paragraphs earlier. It sounds very much like the desire to merge with the universal and infinite will to live that underlies all individual will to live. This is a serious question, which will be raised later. For the moment, I do not want to build an answer into the translation. The German "*eins werden*" means "to agree" or "come to terms with one another." The desire seems to be for a kind of harmony or unity in which there is no conflict, in which the divisions are overcome, and in which the true goals and ideals of others are congruent with my own. The desire to become one with the infinite will to live by leaving my own individualization behind is an example of

the mysticism of the abstract in Schweitzer's eyes. I examine below whether Schweitzer falls into this error.

23. Schweitzer's position here should be contrasted to the abolitionist tendencies in Singer (though his utilitarianism cannot support abolitionism) and Regan's abolitionism concerning medical research on animals. Schweitzer could argue that this difference is a result of the fact that both Singer and Regan base their ethics on abstract principles rather than on a concrete existential foundation.

24. This is the model suggested by Socrates in Plato's *Euthyphro* (6e), but Socrates' practice does not conform to his model. Cf. Evans, "Socratic Ignorance/Socrates Wisdom."

25. Paul Taylor gives a more detailed elaboration of this argument in his denial of human superiority.

26. The English translation uses the word "amusement" here. This is certainly not in any simple sense wrong. "Hunting for amusement" is sometimes the right expression, and I will use it on occasion. I have chosen to translate "*Vergnügung*" as "pleasure" here since I think that it does a better job of conveying the breadth of Schweitzer's rejection of "sport" hunting.

CHAPTER 3

1. For a metaphysical interpretation of Nietzsche's will to power such as Heidegger develops, the parallels would be especially striking—up to a point.

2. "Verily, it is a blessing and not a blasphemy when I teach: 'Over all things stand the heaven Accident, the heaven Innocence, the heaven Chance, the heaven Prankishness'" (Nietzsche, 166).

3. Several interpreters claim that Schweitzer's ethics is based on pity. This interpretation originates with Oskar Kraus, who, writing in 1925, quotes the following passage from *The Philosophy of Civilization* as Schweitzer's summary of his thought:

> *Ethics are pity. All life is suffering. The will to live that has attained to knowledge is therefore seized with deep pity for all creatures. It experiences not only the woe of mankind, but that of all creatures with it. What is called in ordinary ethics "love" is in its real essence pity. In this powerful feeling of pity the will-to-live is diverted from itself. Its purification begins* (Schweitzer 1923, 239; quoted by Kraus, 11–12).

Perhaps under the influence of Kraus, George Seaver also quotes this passage and interprets it as the statement of the "determining factor" of Schweitzer's ethical mysticism (Seaver, 309). But this is a misunderstanding. The passage comes from Schweitzer's chapter on Schopenhauer and Nietzsche, and is a statement of *Schopenhauer's* position, not Schweitzer's own. The penultimate sentence should have tipped them off, because it is an expression of what Schweitzer calls Schopenhauer's attitude of resignation in the sense of world-resignation (Schweitzer 1923, 238).

4. In a note on Paul Taylor's ethic of respect for nature, Ned Hettinger writes that, like Taylor, Schweitzer "seeks to avoid the fundamental natural fact that life thrives at the expense of other life" (Hettinger, 12). Concerning Taylor, see the next section.

5. The quote is accurate. I take it that Harrison means, "assuming that you exclude none of the realities of the natural world and include a meditation on why you hunt."

6. There are of course objections to Levy's argument that attempt to avoid sentimentalism. John Jerome writes, "To be the agent of that transaction [in which an animal is killed]—that moment—carries more responsibility than I think we are willing to take on. It is a moment that, if consciously witnessed, consciously accepted, changes us . . . That is, I'm not so worried about what it does to the animals; I'm worried about what it does to us" (Jerome, 181). But Jerome bears a heavy burden of justification for his claims here, and one has to ask if Jerome's worry is not the result of the very alienation Levy is concerned to counter. Several of the writers in Robert F. Jones's collection *On Killing: Meditations on the Chase* either reject hunting (Gerber, Schreiber) or hunting certain favorite species (Jones), but with the exception of Jerome, they understand this as a very personal decision: "They say the moment [of death] is natural, meant to be, and I know that's often true, but I also know I am not meant to deliver that moment. This stance is not moral, logical, sentimental. It's visceral" (Schreiber, 121).

7. This is not to say that she is committed to his specific form of mysticism, of course.

CHAPTER 4

1. In a letter dated November 6, 2000 Paul Taylor wrote, "Albert Schweitzer's principle of reverence for life did not influence my thinking. I knew of Schweitzer's principle, of course, but only at a shallow level of understanding. I had not read much of his writings, though I admired him as a truly great man. Philosophically, his religiosity was foreign to my feelings about nature and my world view." In his new book *The Green Halo: A Bird's-Eye View of Ecological Ethics*, Erazim Kohák interprets Taylor's theory as a formalization of Schweitzer's ethics of reverence for life (Kohák, xi, 83).

2. This is a very narrow definition of "environmental ethics." As Taylor notes, the distinction between human and environmental ethics is not exhaustive, since it leaves out domestic animals and plants—what Taylor calls the "bioculture" (cf. Taylor 1986, 53–55). It also leaves out cities as human environments. Given my focus on hunting, Taylor's definition need not be objectionable. I do not think that it begs any substantive questions, although at some points it may appear that it does, since by definition human beings are not part of "the natural world." Human participation in the cycles of the natural world will be an issue in my critique of Taylor.

3. In *Respect for Nature*, Taylor is a nominalist about species: the term "species," as well as the term naming a specific species, are "class name[s], and classes themselves have no good of their own, only their members do" (Taylor 1986, 69, ftn. 5). Stephen Jay Gould, who argues for the biological reality of species (cf. Gould), would have a different take on this. See also Michael T. Ghiselin, *Metaphysics and the Origin of Species* and Lewis Petrinovich, *Darwinian Dominion*. James P. Sterba has developed a modified version of Taylor's position that is based on a recognition that species and ecosystems are moral patients. The result is a move from "biocentric individualism" to "biocentric pluralism" (Sterba 1995, 192).

Taylor himself has changed his views. In a letter to Professor Claudia Card written in 1994, Taylor writes that he now follows Lawrence Johnson in extending moral considerability to "whole species, biotic communities, and ecosystems (and perhaps the entire biosphere) as entities to which the concepts of well-being and (objective) interests are applicable" (Taylor 1994).

4. Taylor makes careful distinctions between "intrinsic value," "intrinsic worth," and "inherent worth." For present purposes, I concentrate on "inherent worth." It should be noted that Taylor's conventions have not been generally adopted, and what he calls "inherent worth" others call "intrinsic value."

5. Note that this inclusion of human beings as an "integral part of the natural order" does not fit very well with Taylor's definition of "the natural world" by the exclusion of human intrusion and control (Taylor 1986, 3; cf. ftn. 2 above).

6. It is not clear that Taylor's criticism of Aldo Leopold, whom he considers the inspiration for many holistic views, is well taken. Leopold's thought is more complex than Taylor allows.

7. It is not clear that Taylor would agree with Tom Regan, who characterizes management of animal populations as "environmental fascism."

8. One may wonder whether it makes sense to speak of an organism that is not aware of the world and has no interests at all can have a "point of *view*." The expression is metaphorical, and is to be understood in terms of the notion of having "a constant tendency to protect and maintain the organism's existence" (Taylor 1986, 122).

9. There is more prejudice than fact here. Cf. books such as Guy de la Valdène's *For a Handful of Feathers* (on bobwhite quail) and *Making Game: An Essay on Woodcock*; John J. Mettler, Jr.'s *Wild Turkeys: Hunting and Watching*; David Peterson's *Elkheart*; Richard Nelson's *Heart and Blood: Living with Deer in America*—the list could be continued indefinitely. These authors tell a very different story indeed.

10. While it is clear that for Taylor falconry is morally impermissible, it is not clear what he would say about the virtue of a person who took enormous pleasure in seeing a wild falcon stoop and kill a wild duck. Schweitzer would respond to the spectacle with horror. Taylor would not countenance identifying the predator as the "bad guy," but could he allow for a positive *aesthetic* response to the scene?

CHAPTER 5

1. This point, and the critical remarks that follow, should not be confused with William C. French's assertion that the "strategy" of invoking Taylor's priority principles "oddly suggests that ethical principles ought to be consigned solely to some ideal sphere of pure theory, while our concrete decisions about human action ought to be made outside the sphere of moral review and governed strictly by concerns of power and raw necessity" (French, 40). I think that French misrepresents Taylor's thought here.

2. There is a similarity here to Kant's stoicism. Kant writes, "The inclinations themselves as the sources of needs, however, are so lacking in absolute worth that the universal wish of every rational being must be indeed to free himself completely from them" (Kant 1785, 45). Taylor's environmental stoicism stands Kant's stoicism on its head. In Taylor's version, a genuine respect for nature requires that it be the universal wish of every moral agent (or rational being) that he or she be freed from the natural world as a system of interdependence and interaction, *not*, however, because the natural world is lacking in absolute worth, but precisely because the moral agent is committed to the inherent worth of nature and therefore does not wish to contaminate *it*. I return to this theme of contamination and purity below.

3. Ned Hettinger writes, "Paul Taylor's recent suggestion that the minimization of killing is an appropriate environmental ethical ideal illustrates the saint-like commitment required if we are to minimize harm. Taylor, like Albert Schweitzer before him, seeks to avoid the fundamental natural fact that life thrives at the expense of other life" (Hettinger 12, ftn. 21).

4. I will suggest below that this alternative—the privilege granted by anthropocentric hierarchy versus biocentric equalitarianism in the sense of equal consideration—is not exhaustive.

5. Theodore Vitali makes a similar point (Vitali, 76).

6. This view of ourselves as being outside of the system of interdependence might seem to be the analogue of Kant's postulate of freedom, which Kant thinks is presupposed by morality. But this is a false analogy. The idea of freedom is not inconsistent with the idea of relationships between rational beings.

7. Shepard wrote several critical essays on Schweitzer's ethics of human/animal relations. Cf. "Reverence for Life at Lambaréné" (Shepard, 1958).

8. Shepard was Avery Professor of Natural Philosophy and Human Ecology at Pitzer College. To call his outlook "ecological" does more than verbally distinguish it from Taylor's "biocentric" outlook. When the system of interdependence is taken as basic—basic to meaning—*any* exclusionary "centrism" is inappropriate. Where Taylor's "biocentric outlook" consists of beliefs about "nature," which in Taylor's definition excludes human beings, Shepard's ecological outlook concerns the world of which human beings are a part and human beings as a part of that world. Shepard's outlook would reject Taylor's species impartiality as being hopelessly out of touch with reality, with reality *as lived*, and, more importantly, with the very *meaning* of

life itself. As a result, Shepard approves of attitudes that Taylor would term anthropocentric or expressions of human superiority. Shepard would reject that characterization. I explore this issue in more detail below. (It is noteworthy that the science of ecology plays a role in the development of Taylor's biocentric outlook on nature, but human ecology plays no role at all in *Respect for Nature.*)

9. From an anthropological point of view, this generalization is much too crude. Anthropologist Nurit Bird-David's distinction between a "giving environment" and a "reciprocal environment" in her study of hunter-gatherer societies (Bird-David, 191; cf. Milton 126) would be a start.

10. Put this way, it does not sound so very existential. Taylor's habit of capitalizing such phrases as "Community of Life on Earth" (Taylor 1986, 101) is, however, indicative.

11. Taylor's entire discussion is indebted to the tradition of the harmony or balance of nature, which is put in question by much recent work in ecology. Cf. *Chance and Change*, by William Holland Drury, Jr. and "Disturbing Nature," the final chapter in the second edition of Donald Worster's *Nature's Economy*. It may be that the implications of this new work in ecology have not yet been sorted out clearly, but it seems clear that some of the old paradigms will either have to be rethought or replaced.

12. It is, of course, possible to overdo the analysis of Taylor's rhetoric here, but I think that it is fair to find that Taylor suggests that humans are truly nondestabilizing and (most certainly) nondestructive only when they stay out of the "natural" systems. But this is, of course, not possible, so we should be very careful when we do engage ourselves in these systems. This leads to the priority principles. But it is striking that human beings, who are viewed as part of the system of interdependence, are also viewed as foreign to it, to be allowed in only under the strictest of conditions. The *prima facie* option is always lack of engagement (or preserving, but this *prima facie* obligation faces strict and narrow limits imposed by interdependence). This is an incoherent meaning to give to the fact *and value* of being an "integral part of the system of nature."

13. Similarly, Ann S. Causey writes that Taylor's position here "is partially based on gross anthropomorphism" (Causey, 340). Causey's accusation that Taylor's position "leads to apparent inconsistencies and paradoxes" (*Ibid*, 340) is less convincing, since she leaves out Taylor's rules of nonmaleficence and noninterference, both of which are sufficient to rule out sport hunting.

14. Cf. Thoreau, "Walking"; Turner, "The Abstract Wild: A Rant," "In Wildness Is the Preservation of the World"; Snyder, "The Etiquette of Freedom."

15. A reading of Luke's essay should, I think, be complemented by a reading of Stephen Bodio's essay "A Canvas, Ever Changing," in which he writes, "Hunting Ethics go a lot farther than laws and rules, which determine a minimum or arbitrary set of lines. They involve a lifelong commitment to keeping your eyes open. You can teach a child the rules of gun safety, or how to cast, by rote and repetition. Ethics involve example over a long period, talk, and, above all, thought" (Bodio, 85–86). It is, in other words, about mindfulness.

16. Stephen Bodio writes, "In our society, still enamored of the old Vince Lombardi 'winning is the *only* thing' standard, we might have to uncouple our understanding of field sports from ball games and such to truly understand them, and to come to a concept of honor that does justice to the animals that we hunt and kill" (Bodio, 85).

17. Williams's essay is a marvelous piece of writing straight from the heart, a "rant" in the sense of Jack Turner. For Williams, the "slob" in hunters' self-serving term "slob hunters," is redundant. She goes after hunters' rationalizations with relish, freely using sarcasm as more appropriate to her subject matter than the dispassionate giving of reasons—often to very good effect. Every hunter should read it as a hard look in the mirror. All too many "hunters" will see nothing but themselves peering back at them, whether they are willing to admit it or not.

18. The very idea of any such reciprocal dependence is rejected out of hand by Euthyphro, in Plato's *Euthyphro* (15b), and has generally remained foreign to the Western tradition.

19. Cf. Evans 1984, chapters 4 and 5.

20. One could arrive at a similar, although conceptually distinct, position by arguing that Paul Taylor moves too easily from a question of *value*—the intrinsic value of teleological centers of life—to a moral principle, as if the moral principle could be a simple and direct expression of recognition of the value. This oversimplifies the moral issue.

21. For this reason, I think that Arne Naess's pluralistic and tolerant response to Peter Reed's self-professed "misanthropy" is too weak. It is one thing to ask how we can live worthy lives in the face of "the austere mystery of nature as *Thou*" (Reed, 66). It is something else again to follow Peter Wessel Zapffe in describing humanity as "a sort of evolutionary monster" (*Ibid*). There is at least a tension in claiming that precisely the being that is capable of becoming aware of and appreciating the majesty of nature is *for that very reason* an evolutionary monster. One might even detect a kind of performative contradiction at work when the being capable of recognizing itself to be monstrous declares itself to be, *as such*, a monster. This may not be an appropriate response to a provocation designed to make us reflect on our very reflectivity, but I question both the validity and the effectiveness of trying to undermine human arrogance by insisting on human insignificance (*Ibid*, 56). Is self-hate really the only reason we can find to be good citizens in the biotic community? Indeed, is it really any reason at all?

CHAPTER 6

1. Cf. Part I, n. 11, above.

2. Charles J. List's recent interpretation of Leopold's land ethic in his essay, "Is Hunting a Right Thing?", especially List's emphasis on the land-ethical evaluation of *practices*, can be read as a contemporary attempt to delineate such an ideal of congruency.

3. As we have seen, animal welfare theorists and environmental philosophers have been arguing since the mid-1970s about just what the criterion for moral considerability should be (cf. especially Goodpaster), and this is one way to approach the problem Lopez sketches under the title "new contract." Few have gone as far as José Ortega y Gasset when he asserts that we have obligations to rocks (Ortega y Gasset, 89), and this in a book on the philosophy of hunting!

4. I could just as well write that my position is truly *ecocentric* and egalitarian. In much recent literature in environmental ethics, the term "biocentric" is used for positions that see intrinsic value in individual organisms, in contrast to "ecocentric" views, which see intrinsic value primarily in the whole. My critique of Paul Taylor has the implication that this is a false alternative. Similarly, my position is probably best understood as going beyond anthropocentrism. I have stuck with "biocentric anthropocentrism" precisely because it sounds so paradoxical to those stuck in traditional alternatives.

5. George Sessions writes, "Biocentric egalitarianism is essentially a rejection of human chauvinistic ethical theory . . . Biocentric egalitarianism is essentially a statement of non-anthropocentrism" (Sessions, 5–5a; quoted in Fox, 223). The context is a discussion of Arne Naess's conception of deep ecology, which is characterized by an opposition to focusing on moral "oughts" (cf. Fox, 215–224).

6. The distinction between wise and foolish use would remain, but it would be judged strictly from the perspective of *human* well-being. See the discussion of Murdy in the following pages.

7. As my student Sarah Smith pointed out, if taken even semi-literally, this claim seems committed to the idea of group selection. I have rather taken Murdy's formulations here and in the following quotations as being metaphorical: the species "spider" does not "value" spiders. Indeed, individual spiders don't particularly "value" spiders (newly hatched black widows eat each other voraciously, a survival mechanism for black widow spiders). What is central is reproduction, and the biological point being made is that the species is crucial in understanding evolution (cf. Petrinovich, 12–17, 217; Gould). Lewis Petrinovich offers an extended biologically-based argument for the claim that "the species barrier is an almost insurmountable (and reasonable) one in the moral domain" (Petrinovich, 215).

8. For a broad discussion of the phenomenon of order that is focused on traditional presuppositions such as the anthropocentrism represented here by Kant, cf. Waldenfels.

9. Note that Kant's position here is probably not vulnerable to Carneades' refutation (cf. "Preface" above, ftn. 3).

10. The full first sentence reads, "Not until man places man second, or, to be more precise, not until man accepts his dependency on nature and puts himself in place as part of it, not until then does man put man first" (Iltis, 820).

11. Speaking of a "biocentric anthropocentrism" in this sense must not be confused with Holmes Rolston's construction of what he calls "anthro-

pocentric biocentrism" (Rolston 1988, 77), a position he develops only in order to show what is wrong with it. Rolston's "biocentric anthropocentrism" involves modeling human conduct on animal behavior roughly on the model that since "an alligator will replace a human with an alligator, where possible," it is only natural and right that "one should always replace an alligator with a human, where possible" (*Ibid*).

12. Rolston first develops his principle of nonaddition of suffering, which applies to sentient animals. In his later discussion of nonsentient nature, he notes that the principle of nonaddition of suffering cannot apply where there is no suffering. He then writes, "We can substitute a principle of *the nonloss of goods*" (Rolston 1988, 120). Ned Hettinger correctly points out that this leads to the counterintuitive (on Rolston's own account) result that "the principle protecting plants is stronger than the principles protecting animals" (Hettinger, 7) and proposes that Rolston needs to extend the principle of the nonloss of goods to our relations with animals (*Ibid*, 9).

13. I think that Rolston would agree with this statement. I do not want to suppress artificially the degree to which I am in agreement with his work, but sometimes emphasis is important. Reading an earlier draft of this chapter, Rolston noted:

> *You must eat–*
> *You eat out of necessity–*
> *You surely don't eat out of cultural necessity*
> *What kind of necessity do you eat out of?*
> *Natural necessity.*
> *You are an omnivore–*
> *You don't have to eat meat*
> *You can also & instead eat vegetables*
> *But eating meat is something you are capable of doing, as a result of your evolutionary heritage –*
> *Why respect half of your evolutionary heritage–*
> *I will eat vegetables*
> *I will eat no meat*

14. The relations between the concepts "natural" and "cultural" are confusing to both my introductory students and at times to professional philosophers. Briefly, the "natural/supernatural" is an exclusive distinction—an event is either natural or supernatural (if there be such), but not both. The "natural/cultural" distinction is more complicated. Every cultural event is also, barring supernatural intervention, a natural event, because cultural events (actions, practices, etc.) do not break any laws of nature. Thus, when Moriarty and Woods write that "there is nothing natural about meat eating and hunting in our culture" they have to be understood properly. Their point is not that meat eating is unnatural in any strong sense, but, first, that eating meat is not "natural" in the sense that it is not biologically necessary for human survival and well-being, and, second, that the cultural sphere is the arena in which the normative issues must be decided. If the

"nature versus culture" distinction is intended to be exclusive, then a correspondingly modified sense of "nature" is being invoked.

15. Large-scale huckleberry farming in Montana would likely have an even more harmful effect, since it would destroy the habitat the bears depend on. Thanks to Red Watson for this example.

16. Unfortunately, the review of *Woman the Hunter* by ecofeminist vegetarian Greta Gaard published in *Environmental Ethics* misrepresents Stange's work rather than engaging the issues. Val Plumwood's critique of Stange (Plumwood 2000, 310–314) has to be taken seriously.

17. There are also issues concerning hunting versus gathering, and how they are to be interpreted, here. Stange challenges the interpretation of gathering put forward by Adams and others (cf. Stange, 70f.).

18. Care must be taken in applying such important distinctions as the one Marilyn Frye draws between "the arrogant eye" and "the loving eye" (Frye, 66–76) to practices such as hunting. That much hunting is indeed conducted with an "arrogant eye" is clear—at the very least what Stephen Kellert calls "sport-dominionistic hunting" clearly fits here. To insist *a priori* that all hunting is of necessity an instantiation of the arrogant eye is to turn the "arrogant/loving" distinction into an absolute either/or that can be applied to any individual case without examination of the context.

19. Greta Gaard raises the same charge of "cultural cannibalism" against Karen J. Warren (cf. Gaard 1993, 296).

20. It is also unclear just what Adams means by "gatherer societies" if this is intended to exclude the hunting side of hunter-gatherer societies. As Mary Zeiss Stange points out, there is no clear separation between the activities of hunting and gathering in such societies. To define "gatherer" in terms of an ideal purity that excludes anything Adams would consider to be hunting, and then survey hunter-gatherer cultures to appropriate the pure part, is just as cannibalistic as shallow appropriations of Native-American hunting practices. Ironically, Adams's carnivorous metaphor turns against her.

21. But it is also true that more of us could approach it than do. Forrest Woods, Jr., writes, "I would like to issue a challenge to all ecologically sensitive people who believe in the land ethic: 'Buy some.'" (Woods, 199).

22. Arguments that for environmental reasons one should not eat any meat, even that of wild animals, remind me of a bumper sticker I saw in Germany in the late 1970s. There was a passionate and at times violent public debate concerning nuclear power at the time. The bumper sticker read, "Why Do We Need Nuclear Power Plants? In My House Electricity Comes Out of the Socket in the Wall."

23. Cf. Evans, "Socratic Ignorance—Socratic Wisdom."

24. This emphasis on mindfulness is related to Thomas H. Birch's principle of "universal consideration" in his essay "Moral Considerability and Universal Consideration." My own sense is that Birch fails to locate or situate the moral self in a manner that allows universal consideration to be productive. Alternatively, his account of "deontic experience" lacks a framework, fundamental principle, or basic attitude that can give it moral orientation. Even a completely amoral person can be seriously mindful in many ways. Cf. also

Tim Hayward, "Universal Consideration as a Deontological Principle: A Critique of Birch."

25. Cf. Evans, "Socratic Ignorance—Socratic Wisdom."

26. While I am thus affirming one aspect of Kant's approach to ethical deliberation, I am denying something that Kant thinks is implied by his ethics. In the *Foundations of the Metaphysics of Morals*, Kant writes, "The inclinations themselves as the sources of needs, however, are so lacking in absolute worth that the universal wish of every rational being must be indeed to free himself completely from them" (Kant, 45). This stoicism in Kant's ethics is inconsistent with any environmental ethics that is not strongly anthropocentric. As I point out above, Paul Taylor's ideal of harmony is a form of stoicism within environmental ethics.

27. Kenneth Goodpaster's 1978 essay, "On Being Morally Considerable," helped pose the question of moral considerability for much of the discussion in the 1980s and 1990s. Less attention has been paid to his distinction between moral considerability and moral *significance*. As Goodpaster puts it, a criterion of moral significance has to do with "comparative judgments of moral 'weight' in cases of conflict" (Goodpaster, 309). But Goodpaster does not consider that there might be differences of moral significance that are not matters of more or less, but rather matters of different *kinds* of significance. It is this possibility that I have been exploring.

28. It should be noted that Tom Regan is in agreement at this point.

29. Thomas McGuane inserts the following vignette into his essay "The Heart of the Game":

> *"What did a deer ever do to you?"*
> *"Nothing."*
> *"I'm serious. What do you have to go and kill them for?"*
> *"I can't explain it talking like this."*
> *"Why should they die for you? Would you die for deer?"*
> *"If it came to that"* (McGuane, 237).

30. Paul Taylor has tried to define modes of appropriation that would be appropriate given his concept of "respect for nature." Cf. Taylor, 1990. I argue above that his concept of "respect" is flawed.

CHAPTER 7

1. It should be noted that Ortega is talking about more than trophic levels, since his critique of wildlife photography as a ridiculous mannerism is based on the fact that it treats "the beast as a complete equal" (Ortega y Gasset, 95; cf. 92f.). Ortega's philosophy is a form of vitalism, and while Paul Shepard reads Ortega as challenging the "homocentrism of our traditional philosophy" (Shepard 1972, 18), his philosophy of the hunt is deeply anthropocentric.

2. Sydney Lea writes of seeing a bumper sticker reading "ANIMALS ARE LITTLE PEOPLE IN FUR COATS" just after having to euthanize his old bird dog. His response, in part, is: "Did the driver ahead of me have pets? If so, I pitied them . . . By God, my Annie had never been little people! Just then I felt affected by this slogan as by someone's smirk at a passing funeral cortege. It was—yes—the lack of respect that enraged me" (Lea, 137–138).

Vicki Hearn's descriptions of the difficulties intellectuals, and especially behavioral scientists, have in obedience and riding classes (Hearne, 12) deal with the other extreme of this continuum. Hearne's insistence on the appropriateness and necessity of "the trainers' habit of talking in highly anthropomorphic, morally loaded language" (*Ibid*, 6) points to the moral dimension and moral obligations that are grounded in human-animal *interactions*. But once again, one should not take one part for the whole and model human relations to animals in general on human relations to working animals. It is for this reason that our obligations toward domestic (and especially working) animals are so different from our obligations toward wild animals. This difference is not based on the appropriateness of a morally loaded language for working animals and its inappropriateness for wild animals. It is rather grounded in the essentially different relationships and the appropriately different moral load of the language.

3. And this would even include the practice of catch and release fishing, I suspect. This would be especially important for Meyers, since he is a superb fishing guide, and much of his work involves catch and release fishing.

4. Personal conversation, March 2000.

5. No one who reads Evans's writing can accuse him of leaving the hunter, and the hunter's responsibility, out of the equation. Cf. especially "An Attitude Toward Game" in *The Upland Shooting Life*.

6. Too often any relationship to romanticism is taken to be the kiss of death, mere mention of the connection serving as a deeply telling argument. An example of this can be found in William Cronon's critique of the idea of wilderness as being not only Eurocentric, but also romantic in origin. This is a damning critique only if it is already established that romanticism, and more generally what was then the perspective of specifically Western experience, contained (and contains) no moments of truth. If that is not granted, Cronon's argument largely evaporates. What is wrong with Jones's argument is to be found in its internal contradictions and its assumption that to be a cultural being simply is to be alienated—a legacy of romanticism that is indeed objectionable.

7. My colleague Larry May asks why the interests of the dog count here, since the interests of the bird are not given any weight. This is a good point, especially in light of the fact that hunting dogs are artifacts of human breeding. What has to be added is the beauty and integrity, to someone like George Bird Evans, of the achievements of fine bird dogs. But this brings us back to the fact that for Evans, "shooting is a part of me," and this by itself, as he is aware, does not justify anything. I think that I read Kay Evans' answer as one in which the dogs stand for an entire practice of which they are

an essential moment, and which is essential to the integral being of the dogs themselves.

8. Hunting literature is full of this kind of misuse of the concept of the "contest."

9. The Pastoral Letter itself does not use the term "sacramental commons." The phrase "a sacramental universe" appears in "Renewing the Earth," a pastoral statement of the U. S. Bishops issued in 1991 (cf. Fromherz, 3). The Pastoral Letter can be found at www.columbiariver.org.

10. E. Donnell Thomas speaks of "*The Way of the Outdoors*" (Thomas, 169). Ortega y Gasset uses the word "vocation" in a similar manner, contrasting it with both "occupation" and "diversion" (Ortega y Gasset, 25).

CHAPTER 8

1. James Robb suggests that Byron was reacting to Walton's graphic instructions for using a frog to fish for pike, which culminate in the advice, "Use him as though you loved him, that is, harm him as little as you may possibly, that he may live the longer" (Robb, 22; quoting Walton, 141).

2. Alfred Schutz developed the distinction between the "in-order-to motive" and the "because motive" as an explication of this difference. (Cf. Schutz 1932, 93–105; 1970, 45–52.)

CHAPTER 9

1. PETA spokespersons and websites have used kittens and human beings as the analogues. A PETA web site has the following text under the title "Fishing Hurts":

> *Imagine reaching for an apple on a tree and having your hand suddenly impaled by a metal hook that drags you—the whole weight of your body pulling on that one hand—out of the air and into an atmosphere in which you cannot breathe. That is what fish—who have well-developed pain-receptors—experience when they are hooked for "sport"* (www.Nofishing.net).

2. I take it that this is the reason why speaking of "biting the source of *pain*," in an argument in support of the claim that fish feel pain, is not taken to beg the question.

3. Another example of this line of argument is Neville Gregory's "Can Fish Experience Pain?"

4 Rose warns against widespread misrepresentations of the idea of evolutionary continuity (Rose, personal communication, 3.28.2001). Daniel Dennett's discussion of the idea of "punctuated equilibrium" is helpful in this context (Dennett, 282ff.).

5. Along the same line, my colleague Ilya Farber notes the difference between asking "What makes you think they can't feel pain?" and asking "What makes you think they can?" While at first glance they seem to be two ways of asking the same question, they focus the questioner in different ways. Stoskopf is oriented to the former, Rose to the latter.

6. John S. Kennedy, a sharp critic of anthropomorphism in the scientific study of animal behavior, argues that "anthropomorphic thinking about animals is built into us. We could not abandon it even if we wished to. Besides, we do not wish to." (Kennedy, 5; cf. 31, 159, 167). In Kennedy's scientifically-oriented naturalism, it is difficult to interpret this as more than an evolutionarily useful and to us aesthetically pleasing error.

7. This distinction is derived from Kant's distinction between *phenomena*, or things as they appear to us, and *noumena*, or things as they are in themselves, independent of the conditions under which they appear to us. For Kant, phenomena are subject to the principle of cause and effect, and are in principle part of a deterministic system; noumena are not subject to this condition, and can therefore have an (empirically unverifiable) meaning beyond that of the "ponderable and predictable." Leopold's direct source is not Kant, but the Russian philosopher Piotr Ouspensky's *Tertium Organum*, which Leopold had read by 1922 (cf. Meine, 214; also Flader, 18). Leopold was drawn to Ouspensky's form of life-philosophy ("*Lebensphilosophie,*" cf. Part II) from the perspective of what one might call speculative-organismic ecology (he later largely abandoned the organismic view of biotic systems in favor of the idea of "community"), but not, as far as I can see, to Ouspensky's mysticism. Susan Flader and Baird Callicott note that Leopold never published the 1923 esssay in which he most explicitly appeals to Ouspensky (cf. Leopold 1991, 6). His use of the phenomenon/numenon distinction in *A Sand County Almanac* is clearly aesthetic, not mystical, though I hasten to add that it is in no sense "merely aesthetic." J. Baird Callicott suggests that Leopold's "*numenon*" be read as "aesthetic indicator species" (Callicott 1989c, 242). For Leopold, the *numenon* can be experienced, but not measured or captured by deterministic laws.

8. I have never read an explanation for this locution. It may be that it was originally used to distinguish the cold-water fish we know as "trout" from the warm-water fish that was called "trout" in the late eighteenth-century, namely, largemouth bass (cf. Bartram, 107–108).

9. An outfitter on Vancouver Island offers this excursion, providing the gear and pickup service at the end of the float.

10. In his last book, Harry Middleton wrote,

> *As an average American writer, I have been asked to give and have given one interview.*
> *Here is the whole thing.*
> *INTERVIEWER: "Mr. Middleton, what have fish got to do with it?"*
> *ME: "For Chrissakes, that's what I'd like to know."*
> *What I keep discovering, of course, is that trout and wild rivers,*

> *mountains and sunlight, wind and shadows and the press of*
> *time upon the earth have everything to do with it.*
> *At least for me* (Middleton, 15–16).

11. To be fair, not all hunters will agree with this, and the lines can blur. At the end of his account of a mountain lion hunt, E. Donnell Thomas gives a recipe for "Medaillions of Cougar in Basil Pesto Sauce" (Thomas, 9–10).

12. Lea is in effect extending Ortega y Gasset's analysis of wildlife photography to catch and release fishing.

13. I have argued that this is the essence of the Socratic understanding of excellence or virtue (Evans, 1990).

14. As Jim Harrison notes in his Foreword to the new edition of Harry Middleton's *The Bright Country*, the words

> *Ebb and flow.*
> *The rhythm of things that come and go.*

appear again and again in the book.

15. Thanks to Jack Sadler.

16. The fishing regulations were diabolical. I would have loved a meal of fresh trout, and we were allowed to keep two fish under fourteen inches per day. But small fish do not go up the river to spawn. We never caught a fish under fourteen inches. But there were good reasons for the regulations. Given the fact that the cutthroat trout in Yellowstone Lake are seriously endangered by illegally introduced lake trout, the rumor was that in 2000 only strict catch and release fishing would be allowed. The park now allows only catch and release fishing for cutthroat trout. As I note above, this has brought with it a new set of problems.

17. A lot of scientific research, discussion, teaching, and practice go into learning how best to handle fish that one wishes to release. The state of the art can be studied in Thomas Neil Zacoi's "Catch-and-Release Consequences."

APPENDIX

1. Scully notes in passing that similar sentiments can be found in Islam and Buddhism. He does not mention Hinduism and his dismissal of "the pagan religions" (Scully, 92) is woefully inadequate.

2. Scully does not try to define "sentimentality." Something like "projecting one's own feelings onto animals" would be consistent with his position. He does not mention animal rights advocates who oppose having animals as pets, but he would clearly not agree with them. They might be an example of taking a proper devotion to animals to a sentimental extreme.

3. This would surely change if Scully were to write a general *apologia* for vegetarianism. But the kind of argument he would give there would be very different from the argument in *Dominion.*

4. Subsistence hunting and raising animals for food would presumably be allowed if there were no alternative, but the killing would still be evil.

5. For Scully, the "we" refers to human beings. I am suggesting that it should also include elk and wolves, etc.

6. Of course, this does not really offer a satisfying answer to the problem of the evil of death, since the snake in the Garden of Eden is God's creation and thus presumably truly God's design. Behind the problem of the origin of human sinfulness lies the problem of the snake.

Bibliography

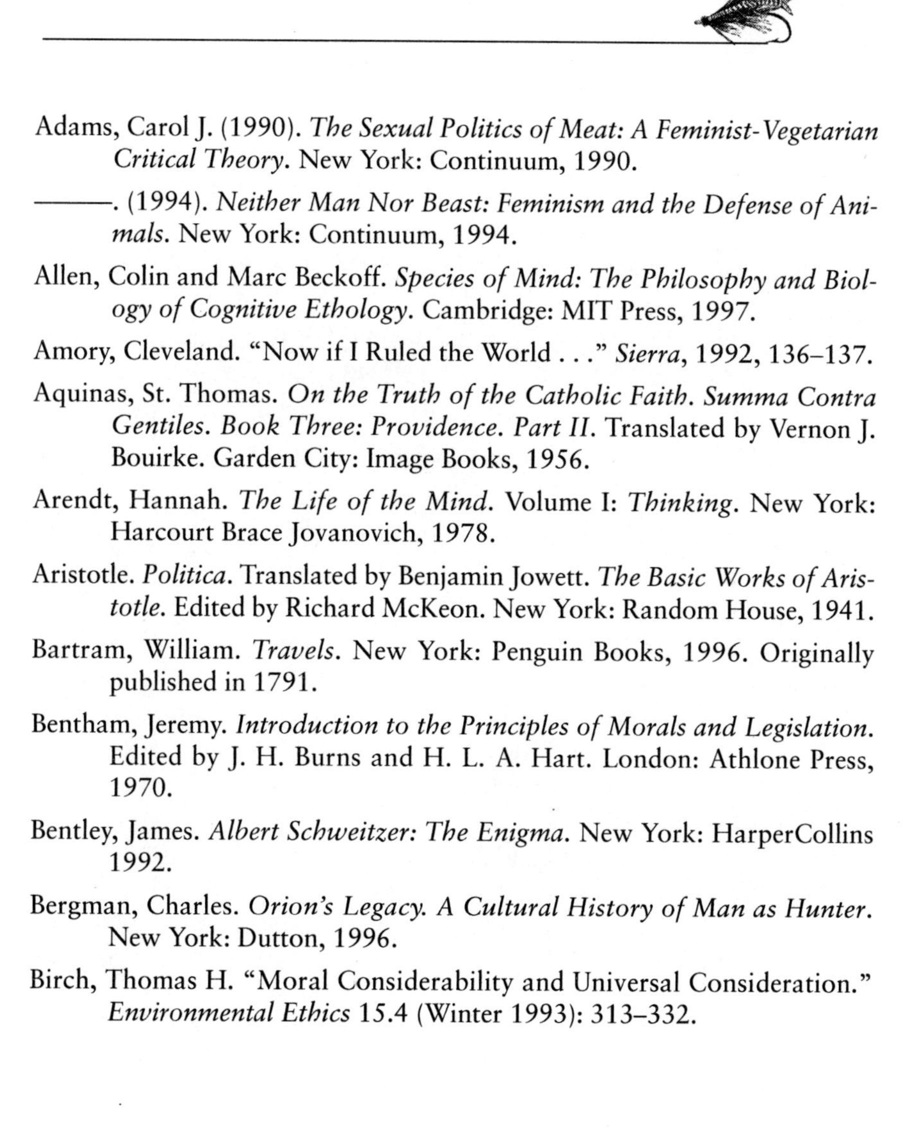

Adams, Carol J. (1990). *The Sexual Politics of Meat: A Feminist-Vegetarian Critical Theory*. New York: Continuum, 1990.

———. (1994). *Neither Man Nor Beast: Feminism and the Defense of Animals*. New York: Continuum, 1994.

Allen, Colin and Marc Beckoff. *Species of Mind: The Philosophy and Biology of Cognitive Ethology*. Cambridge: MIT Press, 1997.

Amory, Cleveland. "Now if I Ruled the World . . ." *Sierra*, 1992, 136–137.

Aquinas, St. Thomas. *On the Truth of the Catholic Faith. Summa Contra Gentiles. Book Three: Providence. Part II*. Translated by Vernon J. Bouirke. Garden City: Image Books, 1956.

Arendt, Hannah. *The Life of the Mind*. Volume I: *Thinking*. New York: Harcourt Brace Jovanovich, 1978.

Aristotle. *Politica*. Translated by Benjamin Jowett. *The Basic Works of Aristotle*. Edited by Richard McKeon. New York: Random House, 1941.

Bartram, William. *Travels*. New York: Penguin Books, 1996. Originally published in 1791.

Bentham, Jeremy. *Introduction to the Principles of Morals and Legislation*. Edited by J. H. Burns and H. L. A. Hart. London: Athlone Press, 1970.

Bentley, James. *Albert Schweitzer: The Enigma*. New York: HarperCollins 1992.

Bergman, Charles. *Orion's Legacy. A Cultural History of Man as Hunter*. New York: Dutton, 1996.

Birch, Thomas H. "Moral Considerability and Universal Consideration." *Environmental Ethics* 15.4 (Winter 1993): 313–332.

Bird-David, Nurit. "The Giving Environment: Another Perspective on the Economic System of Gatherer-Hunters." *Current Anthropology* 31, 2 (1990): 189–196.

Birkner, H.-J. "*Anthropozentrisch.*" *Historisches Wörterbuch der Philosophie.* Volume I. Darmstadt: Wissenschaftliche Buchgesellschaft, 1971, 379.

Bodio, Stephen (1990). *Querencia.* Livingston, MT: Clark City Press, 1990.

———. (1997). "Passions, Gifts, Rages." *A Hunter's Heart: Honest Essays on Blood Sport.* Edited by David Petersen. New York: Henry Holt and Company, 1997, 228–236.

———. (1998). *On the Edge of the Wild.* New York: Lyons Press, 1998.

Bodo, Pete. "Hookless Fly-Fishing: A Humane Advance Over Catch-and-Release." *New York Times*, 7 November 1999, 34.

Bohm, David. *The Van Leer Jerusalem Foundation Series.* New York: Humanities Press, 1973.

Booth, Annie L. and Harvey L. Jacobs. "Ties that Bind: Native American Beliefs as a Foundation for Environmental Consciousness." *Environmental Ethics* 12.1 (Spring 1990): 27–43.

Bourjaily, Vance. *Country Matters.* New York: The Dial Press, 1973.

Brown, Wendy. *Manhood and Politics: A Feminist Reading in Political Theory.* Totowa, NJ: Rowan & Littlefield, 1988.

Buege, Douglas J. "The Ecologically Noble Savage Revisited." *Environmental Ethics* 18.1 (Spring 1996): 71–88.

Campbell, Joseph. *The Power of Myth.* New York: Doubleday, 1988.

Callicott, J. Baird (1989a). "Traditional American Indian and Western European Attitudes Toward Nature: An Overview." *In Defense of the Land Ethic.* Albany: State University of New York Press, 1989, 177–201.

———. (1989b). "American Indian Land Wisdom? Sorting Out the Issues." *In Defense of the Land Ethic.* Albany: State University of New York Press, 1989, 203–219.

———. (1989c). "Leopold's Land Aesthetic." *In Defense of the Land Ethic.* Albany: State University of New York Press, 1989, 239–248.

———. (1989d). "The Conceptual Foundations of the Land Ethic." *In Defense of the Land Ethic.* Albany: State University of New York Press, 1989, 75–100.

———. (1989e). "On the Intrinsic Value of Nonhuman Species." *In Defense of the Land Ethic.* Albany: State University of New York Press, 1989, 129–155.

———. (1992a). "Animal Liberation: A Triangular Affair." *The Animal Rights/Environmental Ethics Debate: The Environmental Perspective*. Edited by Eugene Hargrove. Albany: State University of New York Press, 1992, 37–69. Originally published in 1980.

———. (1992b). "Animal Liberation and Environmental Ethics: Back Together Again." *The Animal Rights/Environmental Ethics Debate: The Environmental Perspective*. Edited by Eugene Hargrove. Albany: State University of New York Press, 1992, 249–261. Originally published in 1988.

———. (1998). "The Wilderness Idea Revisited: The Sustainable Development Alternative." *The Great New Wilderness Debate*. Edited by J. Baird Callicott and Michael P. Nelson. Athens GA: University of Georgia Press, 1998, 337–366. Originally published in 1991.

Callicott, Baird and Michael P. Nelson. *American Indian Environmental Ethics: An Ojibwa Case Study*. Upper Saddle River, NJ: Prentice Hall, 2004.

Carey, Robin. *North Bank*. Corvallis, OR: Oregon State University Press, 1998.

Carr, Dawn. Letter to Philip L. Dubois, President of the University of Wyoming. December 20, 2000. Available on the internet through Virunga News: Something Fishy at the University of Wyoming: http://www.virunga.org/jbin/story/3928.

Causey, Ann S. "On the Morality of Hunting." *Environmental Ethics* 11.4 (Winter 1989): 327–343.

Chase, Alston. *In A Dark Wood: The Fight Over Forests and the Rising Tyranny of Ecology*. Boston: Houghton Mifflin Company, 1995.

Cheney, Jim. "The Neo-Stoicism of Radical Environmentalism." *Environmental Ethics* 11.4 (Winter 1989): 293–325.

Clifton, Chas S. "The Hunter's Eucharist." *A Hunter's Heart: Honest Essays on Blood Sport*. Edited by David Petersen. New York: Henry Holt and Company, 1996, 143–149.

Crisp, Roger. "Evolution and Psychological Unity." *Readings in Animal Cognition*. Edited by Marc Bekoff and Dale Jamieson. Cambridge: MIT Press, 1996, 309–321.

Crist, Eileen. *Images of Animals: Anthropomorphism and Animal Mind*. Philadelphia: Temple University Press, 1999.

Cronon, William. "The Trouble with Wilderness." *The Great New Wilderness Debate*. Edited by J. Baird Callicott and Michael P. Nelson. Athens, GA: University of Georgia Press, 1998, 471–499.

Darwin, Charles. *The Descent of Man and Selection in Relation to Sex*, 2 volumes. Princeton: Princeton University Press, 1981.

de la Valdéne, Guy (1990). *Making Game: An Essay on Woodcock*. Livingston, MT: Clark City Press, 1990.

———. (1995). *For a Handful of Feathers*. New York: Atlantic Monthly Press, 1995.

de Leeuw, A. Dionys. "Contemplating the Interests of Fish: the Angler's Challenge." *Environmental Ethics* 18 (1996): 373–390.

Dennett, Daniel C. *Darwin's Dangerous Idea: Evolution and the Meaning of Life*. New York: Simon & Schuster, 1996.

Devall, Bill and George Sessions. *Deep Ecology. Living As If Nature Mattered*. Layton, UT: Gibbs Smith, Publisher: 1985.

Dobel, Patrick. "The Judeo-Christian Stewardship Attitude to Nature." *Environmental Ethics: Readings in Theory and Application*. Edited by Louis P. Pojman. Belmont, CA: Wadsworth Publishing Company, 1998, 26–30.

Drury, William Holland. *Chance and Change: Ecology for Conservationists*. Berkeley and Los Angeles: University of California Press, 1998.

Dunne, Pete. "Before the Echo." *A Hunter's Heart: Honest Essays on Blood Sport*. Edited by David Petersen. New York: Henry Holt and Company, 1996, 30–34.

Durnil, Gordon K. *The Making of a Conservative Environmentalist*. Bloomington: Indiana University Press, 1995.

Evans, George Bird (1971). *The Upland Shooting Life*. New York: Alfred A. Knopf, 1971.

———. (1982). *An Affair With Grouse*. Old Hemlock, WV: privately published, 1982.

Evans, Joseph Claude (1984). *The Metaphysics of Transcendental Subjectivity: Descartes, Kant, and W. Sellars*. Amsterdam: Verlag B. R. Grüner, 1984.

———. (1990)."Socratic Ignorance—Socratic Wisdom." *The Modern Schoolman*, LXVII (January 1990): 91–109.

———. (1996). "Where is the Lifeworld?" *Issues in Husserl's* Ideas II. Edited by Thomas Nenon and Lester Embree. Dordrecht: Kluwer Academic Publishers, 1996, 57–66.

Fink, Eugen. *Spiel als Weltsymbol*. Stuttgart: W. Kohlhammer Verlag, 1960.

Flader, Susan. *Thinking Like a Mountain: Aldo Leopold and the Evolution of an Ecological Attitude toward Deer, Wolves and Forests*. Columbia: University of Missouri Press, 1974.

French, William C. "Against Biospherical Egalitarianism." *Environmental Ethics* 17.1 (Spring 1995): 39–57.

Fontova, Humberto. "Why We Hunt." *Sierra*, November/December 1990, 54–55.

Fox, Warwick. *Toward a Transpersonal Ecology*. Albany: State University of New York Press, 1995.

Fromherz, Frank. "The Columbia River Watershed: Notes on a Pastoral Letter Emerging," www.columbiariver.org.

Gaard, Greta (1993). "Ecofeminism and Native American Cultures: Pushing the Limits of Cultural Imperialism?" *Ecofeminism: Women, Animals, Nature*. Edited by Greta Gaard. Philadelphia: Temple University Press, 1993, 295–314.

———. (2000). Review of Mary Zeiss Stange, *Woman the Hunter*. *Environmental Ethics* 22.2 (Summer 2000): 203–207.

Gerber, Dan. "A Few Thoughts on Adam's Curse." *On Killing: Meditations on the Chase*. Edited by Robert F. Jones. Guilford, CT: Lyons Press, 2001, 43–47.

Ghiselin, Michael T. *Metaphysics and the Origin of Species*. Albany: State University of New York Press, 1997.

Gierach, John. *Trout Bum: Fly Fishing as a Way of Life*. New York: Simon & Schuster, 1988.

Goethe, Johan Wolfgang von. "An Attempt to Evolve a General Comparative Theory." *Goethe's Botanical Writings*. Translated by Bertha Mueller. Honolulu: University of Hawaii Press, 1952, 81–84.

Goodpaster, Kenneth. "On Being Morally Considerable." *The Journal of Philosophy*, LXXV, 6 June 1978, 308–325.

Gould, Stephen Jay. "What is a Species?" *The Environmental Ethics & Policy Book*. Edited by Donald VanDeVeer and Christine Pierce. Belmont, CA: Wadsworth Publishing Company, 2003, 465–469.

Gregory, Neville. "Can Fish Experience Pain?" Available on the website of the New Zealand Ministry of Agriculture and Forestry: http://maf.govt.nz/mvg/surv26–3/surv26-3-03.htm.

Günzler, Claus. *Albert Schweitzer: Einführung in sein Denken*. Munich: Verlag C. H. Beck, 1996.

Hammond, Bryn. *Halcyon Days: The Nature of Trout Fishing*. Camden, ME: Ragged Mountain Press, 1992.

Harrison, Jim. "Foreword" to Harry Middleton, *The Bright Country*. Boulder, CO: Pruett Publishing Company, 2000, xi–xix.

Haywood, Tim. "Universal Consideration as Deontological Principle: A Critique of Birch." *Environmental Ethics* 18.1 (Spring 1996): 55–63.

Hearne, Vicki. *Adam's Task: Calling Animals by Name*. New York: Vintage Books, 1987, 54).

Henry, James. "Man's Universal Hymn." *The Penguin Book of Victorian Verse*. London: Penguin Books, 1999: 35–39.

Herring, Hal (2000a). "Salting in the Thorofare." *Bugle—Journal of Elk Country* 17.3 (May–June 2000):

——— (2000b). "Money Game." *The Atlantic Monthly*, 6 June 2000, 20–25.

Hettinger, Ned. "Valuing Predation in Rolston's Environmental Ethics: Bambi Lovers versus Tree Huggers." *Environmental Ethics* 16.1 (Spring 1994): 3–20.

Holt, John. "Death on the Musselshell." *On Killing: Meditations on the Chase*. Edited by Robert F. Jones. Guilford, CT: Lyons Press, 2001, 95–101.

Iltis, Hugh H. "Man First? Man Last? The Paradox of Human Ecology." *BioScience* 20.14 (1970): 820.

Jacobs, Wilbur R. "Indians as Ecologists and Other Environmental Themes in American Frontier History." *American Indian Environments: Ecological Issues in Native American History*. Edited by Christopher Vecsey and Robert W. Venables. Syracuse: Syracuse University Press, 1980, 46–64.

Jacobson, Knut A. "The Institutionalization of the Ethics of 'Non-Injury' toward All 'Beings' in Ancient India." *Environmental Ethics* 16.3 (Fall 1994): 287–301.

Jerome, John. "Killing the Neural Pathways." *On Killing: Meditations on the Chase*. Edited by Robert F. Jones. Guilford, CT: Lyons Press, 2001, 177–182.

Jezioro, Frank. *Grouse, Quail and a Splash of Woodcock*. Flemington, WV: Larsen's Outdoor Publishing, 1998.

Johnson, Lawrence. *A Morally Deep World: An Essay on Moral Significance and Environmental Ethics*. Cambridge: Cambridge University Press, 1991.

Jones, Allen Morris. *A Quiet Place of Violence: Hunting and Ethics in the Missouri River Breaks*. Bozeman: Spring Creek Publishing, 1997.

Jones, Robert F. *Dancers in the Sunset Sky: The Musings of a Bird Hunter*. New York: Lyons & Burford, 1996.

Kant, Immanuel. (1775–1780). *Lectures on Ethics*. Translated by Louis Infield. Indianapolis: Hackett Publishing Company, 1980.

———. (1785). *Foundations of the Metaphysics of Morals*. Translated by Lewis White Beck. New York: Macmillan, 1990.

———. (1786). "Speculative Beginning of Human History." *Perpetual Peace and Other Essays on Politics, History, and Morals*. Translated by Ted Humphrey. Indianapolis: Hackett Publishing Company, 1983, 49–60.

Kellert, Stephen R. (1978a). "Attitudes and Characteristics of Hunters and Antihunters." *Transactions of the North American Wildlife and Natural Resources Conference*, Volume 43. Washington, D.C.: Wildlife Management Institute, 1978: 412–413.

———. (1978b). *Policy Implications of a National Study of American Attitudes and Behavioral Relations to Animals*. Washington, D.C.: U.S. Government Printing Office, 1978.

Kennedy, John S. *The New Anthropomorphism*. Cambridge: Cambridge University Press, 1992.

Kerasote, Ted. (1994). *Bloodties. Nature, Culture, and the Hunt*. New York: Kodansha International, 1994.

———. (1997). "Catch and Deny." *Orion* 16.1 (Winter 1997): 24–27.

———. (1999). "The New Hunt." *Sports Afield* 220. 5 (November 1999): 78–82.

Kierkegaard, Søren 1843a). *Either/Or*. Edited and translated by Howard V. Hong and Edna H. Hong. Princeton: Princeton University Press, 1987.

———. (1843b). *Fear and Trembling*. Edited and translated by Howard V. Hong and Edna H. Hong. Princeton: Princeton University Press, 1983.

———. (1846). *Concluding Unscientific Postscript*. Edited and translated by Howard V. Hong and Edna H. Hong. Princeton: Princeton University Press, 1992.

Kohák, Erazim. *The Green Halo: A Bird's-Eye View of Ecological Ethics*. Chicago: Open Court, 2000.

Kraus, Oskar. *Albert Schweitzer: His Work and His Philosophy*. London: Adam & Charles Black, 1944. The original German version was written in 1925.

Krutch, Joseph Wood. "Reverence for Life." *The Best Nature Writing of Joseph Wood Krutch*. New York: William Morrow, 1969.

Lea, Sidney. *Hunting the Whole Way Home*. New York: The Lyons Press, 2002

Leopold, Aldo. (1949). *A Sand County Almanac, And Sketches Here and There*. New York: Oxford University Press, 1949, 1987.

———. (1991). *The River of the Mother of God and Other Essays*. Edited by Susan L. Flader and J. Baird Callicott. Madison: University of Wisconsin Press, 1991.

Levinas, Emmanuel. (1969). *Totality and Infinity*. Translated by Alphonso Lingis. Pittsburgh: Duquesne University Press, 1969.

———. (1981). *Otherwise Than Being Or Beyond Essence*. Translated by Alphonso Lingis. The Hague: Martinus Nijhoff Publishers, 1981.

Levy, Buddy. *Echoes on Rimrock. In Pursuit of the Chukar Partridge*. Boulder: Pruett Publishing Company, 1998.

List, Charles. (1997). "Is Hunting a Right Thing?" *Environmental Ethics* 19.4 (Winter 1997): 405–416.

———. (2000). "On the Definition and Moral Appraisability of Sport Hunting." Presented at the Eastern Division meeting of the American Philosophical Association in New York, December 2000.

Long, A. A. and D. N. Sedley, eds. *The Hellenistic Philosophers*, Volume 1. Cambridge: Cambridge University Press, 1987.

Lopez, Barry. (1987). *Arctic Dreams*. New York: Bantam Books.

———. (1991). "Renegotiating the Contracts." In *This Incomperable Lande*. Edited by Thomas J. Lyon. New York: Penguin Books, 381–388.

Luce, A. A. *Fishing and Thinking*. Camden, ME: Ragged Mountain Press, 1993. Originally published in 1959.

Luke, Brian. (1996). "Justice, Caring, and Animal Liberation." *Beyond Animal Rights*. Edited by Josephine Donovan and Carol J. Adams. New York: Continuum, 77–102.

———. (1997). "A Critical Analysis of Hunters' Ethics." *Environmental Ethics* 19.1 (Spring 1997): 25–44.

Maclean, Norman. *A River Runs Through It*. Chicago: University of Chicago Press, 1976.

McCully, C. B. *The Other Side of the Stream: the Mysteries of Fly-Fishing*. Shrewsbury, England: Swan Hill Press, 1998.

McGuane, Thomas. (1982). "The Heart of the Game." *An Outside Chance: Essays on Sport*. New York: Penguin Books, 227–243.

———. (2000). *Upstream: Fly Fishing in the American West*. Photographs by Charles Lindsay. Millerton, NY: Aperture.

Meine, Curt. *Aldo Leopold: His Life and Work*. Madison: University of Wisconsin Press, 1988.

Mettler, John J. *Wild Turkeys: Hunting and Watching*. Pownal, VT: Storey Books, 1998.

Meyers, Stephen J. "Of Dogs and Men." (Unpublished manuscript.)

Middleton, Harry. *The Bright Country. A Fisherman's Return to Trout, Wild Water, and Himself*. New York: Simon & Schuster, 1993. Republished by Pruett Publishing Company, Boulder, CO, 2000.

Milton, Kay. *Environmentalism and Cultural Theory: Exploring the Role of Anthropology in Environmental Discourse*. London and New York: Routledge, 1996.

Moriarty, Paul Veatch and Mark Woods. "Hunting ≠ Predation." *Environmental Ethics* 18.4 (Winter 1997): 391–404.

Murdy, William H. "Anthropocentrism: A Modern Version." *Environmental Ethics: Divergence and Convergence*. Edited by Richard G. Botzler and Susan J. Armstrong. Boston: McGraw Hill, 1998, 316–323. Originally published in *Science* 187 (1985): 1168–1172.

Naess, Arne. (1988). "Self-Realization: An Ecological Approach to Being in the World." *Thinking Like a Mountain*. Edited by John Seed, Joanna Macy, Pat Fleming, and Arne Naess. Philadelphia: New Society Publishers, 19–30.

———. (1990). "*Man Apart* and Deep Ecology: A Reply to Reed." *Environmental Ethics* 12.2 (Summer, 1990): 185–192.

Nelson, Richard. (1969). *Hunters of the Northern Ice*. Chicago: University of Chicago Press.

———. (1983). *Make Prayers to the Raven*. Chicago: University of Chicago Press.

———. (1991). *The Island Within*. New York: Vintage Books.

———. (1994). "Life-Ways of the Hunter." (Interview with Richard Nelson). *Talking on the Water: Conversations About Nature and Creativity*. Written by Jonathan White. San Francisco: Sierra Club Books, 79–97.

———. (1997). *Heart and Blood: Living with Deer in America*. New York: Alfred A. Knopf, 1997.

Nietzsche, Friedrich. *Thus Spoke Zarathustra: A Book for None and All.* Translated by Walter Kaufmann. New York: Penguin Books, 1978. *Also Sprach Zarathustra. Ein Buch für Alle und Keinen.* In *Sämtliche Werke*, Volume 4. Edited by Giorgio Colli and Mazino Montinari. Berlin: Walter de Gruyter & Co., 1980.

Norton, Bryon. Review of Paul Taylor, *Respect for Nature. Environmental Ethics* 9.3 (Fall 1987): 261–267.

Ortega y Gasset, José. *Meditations on Hunting.* Translated by Howard B. Wescott. Bozeman, MT: Wilderness Adventures Press, 1995. Translation originally published in New York by Charles Scribner's Sons in 1972.

Owens, Louis. "The Hunter's Dance." *On Killing: Meditations on the Chase.* Edited by Robert F. Jones. Guilford, CT: Lyons Press, 2001, 183–194.

Petersen, David. *Elkheart: A Personal Tribute to Wapiti and Their World.* Boulder: Johnson Books, 1998.

Petrinovich, Lewis. *Darwinian Dominion: Animal Welfare and Human Interests.* Cambridge: MIT Press, 1999.

Pflug, G. "*Lebensphilosophie.*" *Historisches Wörterbuch der Philosophie,* Volume 5. Darmstadt: Wissenschaftliche Buchgesellschaft, 1980, 135–139.

Pianka, Eric R. *Evolutionary Ecology.* New York: HarperCollins, 1994.

Plato. *Euthyphro. Gorgias. Apology.* The Dialogues of Plato, Volume I. Translated by R. E. Allen. New Haven: Yale University Press, 1984.

Plumwood, Val. (1993). *Feminism and the Mastery of Nature.* London: Routledge.

———. (2000). "Integrating Ethical Frameworks for Animals, Humans and Nature: A Critical Feminist Eco-Socialist Analysis." *Ethics and the Environment* 5.2: 285–322.

Posewitz, Jim. *Beyond Fair Chase: The Ethic and Tradition of Hunting.* Helena and Billings, MT: Falcon Press Publishing Company, 1994.

Proper, Datus. *Pheasants of the Mind: A Hunter's Search for a Mythic Bird.* Bozeman, MT: Wilderness Adventures Press, 1990.

Quammen, David. "Synecdoche and the Trout." *Wild Thoughts from Wild Places.* New York: Scribner, 1998, 19–26.

Raines, Howell. *Fly Fishing Through the Midlife Crisis.* New York: William Morrow and Company, 1993.

Reed, Peter. "Man Apart: An Alternative to the Self-Realization Approach." *Environmental Ethics* 11.1 (Spring 1989): 53–69.

Regan, Tom. (1982). "The Nature and Possibility of an Environmental Ethic." *All That Dwell Therein*. Berkeley and Los Angeles: University of California Press.

———. (1983). *The Case for Animal Rights*. Berkeley and Los Angeles: University of California Press.

———. (1985). "The Case for Animal Rights." *The Environmental Ethics & Policy Book*. Edited by Donald VanDeVeer and Christine Pierce. Belmont, CA: Wadsworth Publishing Company, 1998, 102–109. Originally published in 1985.

Rilke, Rainer Maria. "*Archaïscher Torso Apollos.*" *Werke in Sechs Bänden*, Volume I. Frankfurt am Main: Insel Verlag, 1980, 313. English translation by J. B. Leishman: "Archaic Torso of Apollo." *Rainer Maria Rilke: Prose and Poetry.* Edited by Egon Schwarz. New York: Continuum, 1984, 189.

Robb, James. *Notable Angling Literature*. Ashburton, England: The Flyfisher's Classic Library, 1998. Originally published in 1947.

Robbins, Jim. (2000a). "Holy Water." *High Country News*. 11 September 2000, 1, 8–12.

———. (2000b). "God and Nature: Saving Souls and Salmon." *New York Times*, 22 October 2000.

Rolston III, Holmes. (1988). *Environmental Ethics: Values in and Duties to the Natural World.* Philadelphia: Temple University Press.

———. (1991a). "Environmental Ethics: Values in and Duties to the Natural World." In *Ecology, Economics, Ethics: The Broken Circle.* Edited by F. Herbert Bormann and Stephen R. Kellert. New Haven: Yale University Press, 73–82.

———. (1991b). "The Wilderness Idea Reaffirmed." *The Great New Wilderness Debate*. Edited by J. Baird Callicott and Michael P. Nelson. Athens, GA: University of Georgia Press, 1998, 367–394. Originally published in 1991.

———. (1994). *Conserving Natural Value*. New York: Columbia University Press.

———. (1998). "Feeding People Versus Saving Nature?" *The Environmental Ethics and Policy Book*. Edited by Donald VanDeVeer and Christine Pierce. Belmont, CA: Wadsworth Publishing Company, 409–420.

Rose, James D. "Do Fish Feel Pain?" *In-Fisherman* (December–January 1999–2000): 38–42.

———. (2002). "The Neurobehavioral Nature of Fishes and the Question of Awareness and Pain." *Reviews in Fisheries Science* 10.1, 1–38.

Sapontzis, Steve S. "Predation." *Ethics and Animals* 5 (June 1984): 27–36.

Seaver, George. *Albert Schweitzer: The Man and His Mind.* London: Adam & Charles Black, 1969, 1947.

Schreiber, Le Anne. "Predilections." *On Killing: Meditations on the Chase.* Edited by Robert F. Jones. Guilford, CT: The Lyons Press, 2001, 115–121.

Schutz, Alfred. (1932). *Der sinnhafte Aufbau der sozialen Welt.* Vienna: Springer Verlag, 1932, 1960. English translation by George Walsh and Frederick Lehnert. *The Phenomenology of the Social World.* Evanston: Northwestern University Press, 1967.

———. (1970). *Reflections on the Problem of Relevance.* New Haven: Yale University Press, 1970.

Schweitzer, Albert. (1919). *A Place for Revelation: Sermons on Reverence for Life.* Translated by David Larrimore Holland. New York: Macmillan, 1988. *Was Sollen Wir Tun? 12 Predigten über ethische Probleme.* Heidelberg: Verlag Lambert Schneider, 1974. Sermons on reverence for life delivered in 1919.

———. (1923). *The Philosophy of Civilization.* Translated by C. T. Campion. Buffalo: Prometheus Books, 1987. *Kulturphilosophie. Gesammelte Werke,* Volume 2. Munich: Verlag C. H. Beck, 1973. First published in 1923.

———. (1931). *Out of My Life and Thought.* Translated by Ann B. Lemke. New York: Henry Holt and Company, 1990. First published in 1931.

———. (1937). "*Afrikanische Jagdgeschichten.*" *Gesammelte Werke,* Volume 5. Munich: Verlag C. H. Beck, 1973. Originally published in 1937.

———. (1949). *Memoirs of Childhood and Youth.* Translated by C. T. Campion. New York: Macmillan.

———. (1950). *The Animal World of Albert Schweitzer.* Translated and edited by Charles R. Joy. Hopewell, NJ: The Ecco Press.

———. (1952a). *Indian Thought and Its Development.* Translated by Mrs. Charles E. B. Russell. Boston: Beacon Press, 1952. *Die Weltanschauung der Indischen Denker. Gesammelte Werke,* Volume 2. Munich: Verlag C. H. Beck, 1973, 421–662.

———. (1952b). "The Problem of Ethics in the Advancement of Human Consciousness." *The Teaching of Reverence for Life*. Translated by Richard and Clara Winston. New York: Holt, Rinehart and Winston, 1965, 9–30. *Das Problem der Ethik in der Höherentwicklung des Menschen Denkens.*" *Gesammelte Werke*, Volume 5. Munich: Verlag C. H. Beck, 1973, 143–159. Originally published in 1952.

———. (1963). "*Die Entstehung der Lehre der Ehrfurcht vor dem Leben und ihre Bedeutung für unsere Kultur.*" *Gesammelte Werke*, Volume 5. Munich: Verlag C. H. Beck, 1973. Written in 1963.

Scully, Matthew. *Dominion. The Power of Man, the Suffering of Animals, and the Call to Mercy*. New York: St. Martin's Press, 2002.

Seccombe, Joseph. *A Discourse Utter'd in Part at Ammauskeeg-Falls in the Fishing-Season • 1739*. Barre, MA: Barre Publishers, 1971.

Sessions, George. *Ecophilosophy III* (newsletter). 1981.

Shepard, Paul. (1958). "Reverence for Life at Lambaréné." *Encounters With Nature*. Washington, D.C.: Island Press, 1999, 56–66. Originally published in *Landscape* (Winter 1958–1959): 26–29.

———. (1972). "Introduction" to José Ortega y Gassett, *Meditations on Hunting*, 15–19.

———. (1974). "Animal Rights and Human Rites." *Traces of an Omnivore*. Washington, D.C.: Island Press, 1996, 11–26. Originally published in *The North American Review* (Winter 1974).

———. (1991). "To Hunt." Edited by Florence Shepard. *The Bugle* (July/August 1999): 45–46. Originally published as "Searching Out Kindred Spirits" in the Summer 1991 issue of *Parabola*.

———. (1996). *The Others: How Animals Made Us Human*. Washington, D.C.: Island Press.

———. (1998). *Coming Home to the Pleistocene*. Edited by Florence R. Shepard. Washington, D.C.: Island Press.

Singer, Peter. (1973). "Animal Liberation." *The Environmental Ethics & Policy Book*. Edited by Donald VanDeVeer and Christine Pierce. Belmont, CA: Wadsworth Publishing Company, 1998, 95–102. Originally published in *The New York Review of Books* in 1973.

———. (1990). *Animal Liberation*. New York: Avon Books.

———. (1993). *Practical Ethics*. New York: Cambridge University Press.

Snyder, Gary. *The Practice of the Wild*. San Francisco: North Point Press, 1990.

Sorabji, Richard. *Animal Minds & Human Morals. The Origins of the Western Debate*. Ithaca: Cornell University Press, 1993.

Stange, Mary Zeiss. *Woman the Hunter*. Boston: Beacon Press, 1997.

Stephens, William O. "Stoic Naturalism, Rationalism, and Ecology." *Environmental Ethics* 16.3 (Fall 1994): 275–286.

Sterba, James P. (1995). "From Biocentric Individualism to Biocentric Pluralism." *Environmental Ethics* 17.2 (Summer 1995): 191–207.

———. (1998). "A Biocentrist Strikes Back." *Environmental Ethics* 20.4 (Winter 1998): 361–376.

Stoskopf, Michael K. "Pain and Analgesia in Birds, Reptiles, Amphibians, and Fish." *Investigative Ophthalmology and Visual Science* 35.2 (February 1994): 775–780.

Strong, David. *Crazy Mountains: Learning From Wilderness to Weigh Technology*. Albany: State University of New York Press, 1995.

Swan, James A. *In Defense of Hunting*. New York: HarperCollins, 1995.

Talbot, Ret. "Catch-and-Kill." *Fly Fisherman* 32.4 (May 2001): 14–18.

Taylor, Charles. *The Ethics of Authenticity*. Cambridge: Harvard University Press, 1991.

Taylor, Paul W. (1981). "The Ethics of Respect for Nature." *Environmental Ethics* 3 (Fall 1981): 197–218.

———. (1986). *Respect for Nature: A Theory of Environmental Ethics*. Princeton: Princeton University Press.

———. (1990). "Some Notes on the Ethics of the Bioculture." Letter to Professor Gary Comstock. Quoted by permission of Paul Taylor.

———. (1994). Letter to Professor Claudia Card. Quoted by permission of Paul Taylor.

———. (2000a). "Notes on the work in progress of Professor Claude Evans." Handwritten manuscript. September 2000. Quoted by permission of Paul Taylor.

———. (2000b). Letter to J. Claude Evans. November 6, 2000. Quoted by permission of Paul Taylor.

———. (2001). Letter to J. Claude Evans. February 1, 2001. Quoted by permission of Paul Taylor.

Thomas, Keith. *Man and the Natural World. A History of the Modern Sensibility*. New York: Pantheon Books, 1983.

Thoreau, Henry David. *The Selected Works of Thoreau*. Boston: Houghton Mifflin Company, 1975.

Thomas, E. Donnell. *To All Things a Season*. Gallatin Gateway, MT: Wilderness Adventures Press, 1997.

Tullis, Larry. *Fly Fishing for Trout. Small Fly Techniques*. Birmingham, AL: Odysseus Editions, 1993.

Turner, Jack. *The Abstract Wild*. Tucson: University of Arizona Press, 1996.

Varner, Gary. *In Nature's Interests? Interests, Animal Rights, and Environmental Ethics*. New York: Oxford University Press, 1998.

Vecsey, Christopher. "American Indian Environmental Religions." *American Indian Environments: Ecological Issues in Native American History*. Edited by Christopher Vecsey and Robert W. Venables. Syracuse: Syracuse University Press, 1980, 1–37.

Vitali, Theodore. (1990). "Sport Hunting: Moral or Immoral?" *Environmental Ethics* 12.1 (October 1990): 69–82.

———. (1996). "Why Fair Chase?" *Fair Chase*. Part I: Spring 1996, 20–24. Part II: Summer 1996, 20–23.

Waldenfels, Bernhard. *Order in the Twilight*. Translated by David J. Parent. Athens, OH: Ohio University Press, 1996.

Walton, Isaak. *The Compleat Angler*. Hopewell, NJ: Ecco Press, 1995, 1653.

Warren, Karen J. "The Power and Promise of Ecological Feminism." *Environmental Ethics* 12 (Summer 1990): 125–146.

White, Jr., Lynn. "The Historical Roots of Our Ecologic Crisis." *Science* 155, No. 3767 (March 1967): 1203–1207.

Wilder, Kathryn. "So Close to the Wild." *Bugle* 17.5 (September–October 2000): 87–94.

Williams, Joy. "The Killing Game." *Women on Hunting*. Edited by Pam Houston. Hopewill, NJ: Ecco Press, 1995, 248–265.

Williams, Terry Tempest. "Deerskin." In *A Hunter's Heart: Honest Essays on Blood Sport*. Edited by David Peterson. New York: Henry Holt and Company, 1996, 51–57.

Wood, Jr., Forest. *The Delights and Dilemmas of Hunting*. Lanham, MD: University Press of America, 1997.

Worster, Donald. *Nature's Economy: A History of Ecological Ideas*. New York: Cambridge University Press, 1997.

Wulff, Lee. *Lee Wulff's Handbook of Freshwater Fishing*. New York: Frederick A. Stokes Co., 1939.

Yatzeck, Richard. *Hunting the Edges.* Madison: University of Wisconsin Press, 1999.

Zacoi, Thomas Neil. "Catch-and-Release Consequences: Trout Die If You Don't Release Them Properly." *Flyfisherman* 32:2 (February 2001): 11–12, 15.

INDEX